Design Quality and the Private Finance Initiative

The Royal Fine Art Commission Seminar
25 March 1997

Chairman: Bryan Jefferson CB CBE

Thomas Telford

Contents

Introduction

Lord St John of Fawsley
Chairman, Royal Fine Art Commission

I am delighted to welcome you to the Royal Fine Art Commission's latest seminar. Most of the Commission's resources are devoted to considering casework and seeking to improve individual development schemes. But a vital part of our rôle is to reflect on wider strategic issues; seminars, and the publications which follow them, allow us to do so critically and in some depth.

Today we have assembled some highly distinguished speakers to discuss a matter of real and topical importance. As many of you will know, the Commission has not shrunk from expressing its reservations about PFI. Because the timetable is tight and discussion important, I will not rehearse them here at length. Suffice to say that from the point of view of design quality, there are inherent dangers in concentrating responsibility for designing, building, financing and operating a building in one service provider. Of course the architect will be an integral member of the provider's team, but he will most probably be a low-ranking one. Therein lies a serious danger: if the relationship between the client and architect is weakened, if the brief is filtered through an intermediary whose interests lie elsewhere, then the prospects for high-quality architecture must surely be reduced.

Now that PFI projects are coming on stream, as they are doing gradually, we at the Commission are seeing some of our fears realised. In recent weeks we have been asked to comment on hospital designs which fall far short of the required standard. The designs for the new Inland Revenue building at Bootle are equally lacking in promise. I fear that the Commission will become increasingly familiar with the problems of carelessly-handled elevations, intensive ancillary development, over-provision of car-parking and token landscaping. These problems are hardly unique to PFI schemes, neither do they inevitably result from PFI, but the mechanics of PFI may make their occurrence more likely.

But let us try to be optimistic. We still have the opportunity to shape the architectural legacy of PFI. Valuable lessons can be learnt from the two case studies being presented today – the Central Middlesex Hospital, and the

Royal Armouries Museum in Leeds. PFI comes in many guises, but if the variant is carefully chosen and properly applied, it is perfectly possible for good design to be delivered. For that to happen, we need procurers who demand quality and select their preferred bidder with quality as well as cost in mind – procurers who understand the simple economic truth that design quality equals value for money.

One reason for optimism is that the responsible agencies are well aware of the Commission's concerns and are acting to address them. There is the prospect of design guidance from the Treasury's Private Finance Panel. Mr Horam's NHS Quality Initiative allows us to hope that many of the fine words in *Better by Design* will be translated into action.[1] And the Department of National Heritage, together with the Private Finance Panel, has set up a steering group to advance the cause of good PFI architecture. All this industry must, of course, generate more light than heat, or else it will be largely futile. It is a huge task to create, by a mixture of education, persuasion and prescription, a body of informed and motivated public clients which can both discern and demand quality. As a starting point, might we not nudge clients in the right direction by giving them access to expert design advice in the initial stages of a project? By that simple expedient, we could do much to lift the standard of PFI architecture.

A further reason for optimism is that Bryan Jefferson is closely involved with current attempts to improve the design quality of PFI schemes, particularly in his capacity as chairman of the DNH/Private Finance Panel steering group. The Jefferson/Delafons Report of 1991 predates PFI but remains an excellent analysis of the challenges faced by those who procure and design Government buildings. This seminar will benefit considerably from his experience and wisdom, and it gives me great pleasure to invite him to take the chair.

Bryan Jefferson

Over recent years the Commission has arranged a selection of seminars on relevant and timely subjects. They have shown an unerring ability to come up at the right time with the right questions. Those seminars have been widely appreciated, both by the delegates attending and by those who have read the published proceedings.

Today promises to maintain that record. We have a wide range of speakers who, between them, will cover the aims and intentions behind the Private Finance Initiative. Many have direct experience of working within that régime.

Finally, but certainly not least, we have a distinguished and authoritative list of delegates here today whose views will, I hope, be heard.

---

[1] *Better by Design*, NHS Estates, HMSO 1994

Setting the Scene

Deborah Bartlett
Private Finance Panel, HM Treasury

This is a brief summary which aims to remind us of the policy context of today's discussion. I will run quickly through the background to PFI, looking at when, how and why it was developed, its mechanics and its aims and aspirations.

Lying behind Norman Lamont's launch in 1992 of the Private Finance Initiative was some intensive, if rapid, thinking by senior Treasury officials, many of whom had been deeply involved in conceiving and implementing the Government's privatisation programme. Spurring on this thinking were the economic and political realities of Britain in the early 1990s. I am sure we can all recall that we were in a state of deep and deepening recession; the Public Sector Borrowing Requirement was persistently and unacceptably high; there was continuing decline in the quality of public service delivery and, worse, in the condition of the buildings and infrastructure from which public services were delivered.

Attempts up to that point to draw private investment into partnership delivery of public infrastructure and services had largely collapsed, I think because of the interpretation of the Treasury's Ryrie Rules, which tended to over-emphasise the merits of public sector comparators because of the relative cheapness of public sector borrowing. There was, of course, immense political pressure to continue the downward trend in direct taxation. The forward look, therefore, was that the public pot was going to get smaller rather than larger.

Concurrently, however, the process of privatising state-run enterprises which were already operating as self-contained businesses or as public sector monopolies was rolling steadily on. For the second-order infrastructure and services – the more fragmented businesses such as health and education, or those always likely to need substantial subsidy, or those whose privatisation might just be beyond the pale for the British electorate – privatisation was not perceived to be a workable solution.

And so began the exercise of harnessing the best aspects of privatisation and moulding them into a process which could deliver much-needed investment into these sorts of public services and their related infrastructure without increasing the PSBR.

In short, the aims were to introduce into the delivery of public service the dynamic of privatisation; to bring in private sector management skills and

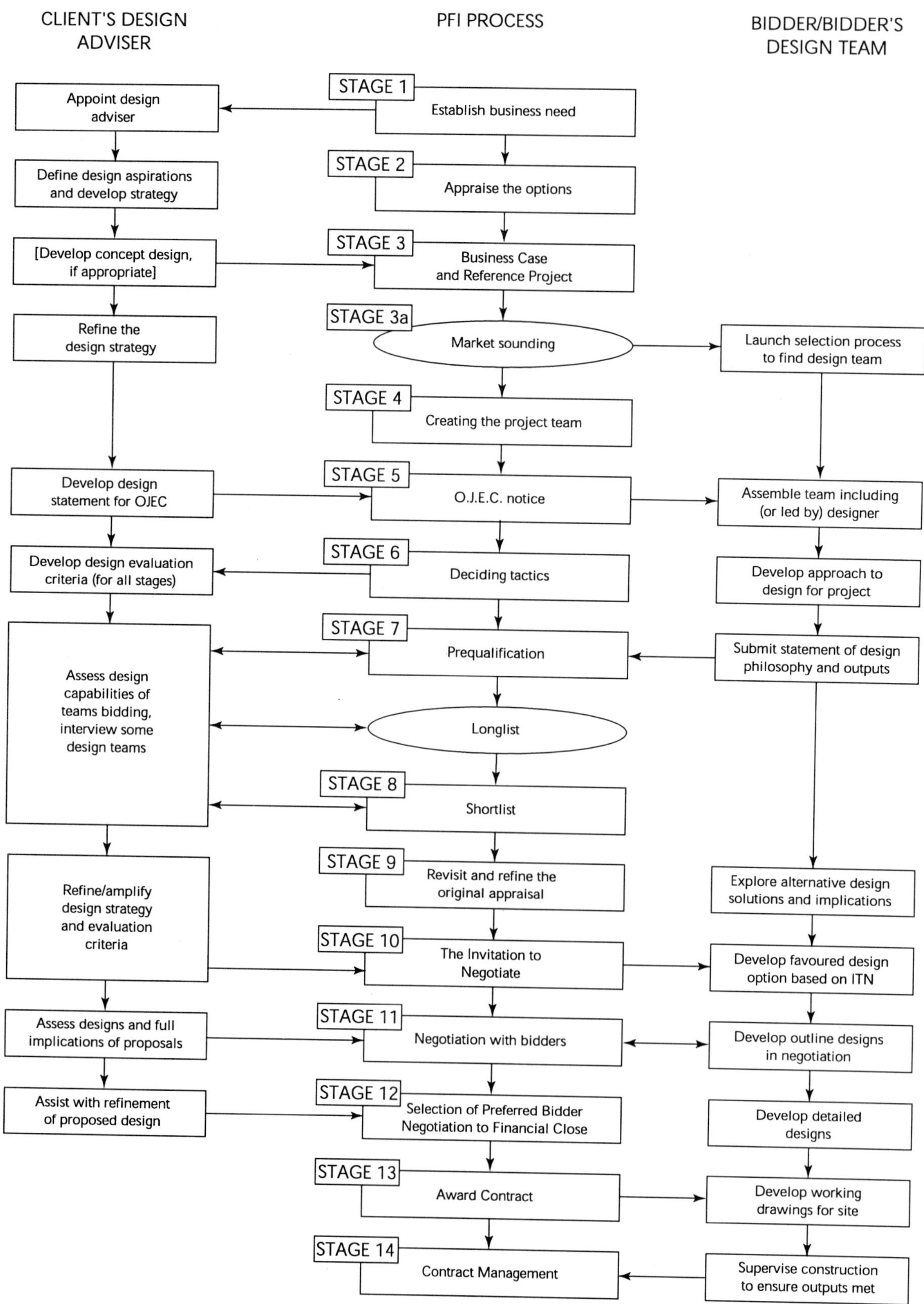

The PFI procurement process (Private Finance Panel Executive)

techniques to replace those of the public sector (which sounds critical of the public sector, but I think it was a recognition of enforced impotence as much as anything else); to introduce accountability into public service delivery, by introducing the concept of risk – risk identification, evaluation and management. And then to go one step further and transfer identified and quantified risks to those best placed to manage them – which, in the case of design and construction, would be private sector consortia. Finally, and most importantly, all this would extract better value for the public from each pound of public money and thereby secure better and greater service provision from a limited pot.

So, without going into all the details, the PFI was born. Misleadingly named, perhaps, for while it depends crucially on the access that private enterprise has to sources of finance and funding mechanisms which are out of reach to the public sector, it is fundamentally a procurement method by which the public sector can set the parameters, including design considerations, for service delivery. It can then monitor that delivery, while being relieved of many of the conventional associated burdens such as heavy capital investment and asset ownership and maintenance – generally in the context of very uncertain and ever-changing, usually diminishing, budgets. The public sector outlay is for services delivered: pre-agreed service payments on delivery of whatever the required services are, to pre-agreed standards.

Among the implications of all of this is that the procuring teams within Government departments – the NHS, local government, or wherever it may be – must invest a lot of careful thought in the initial definition of the asset-based service they are about to procure, before the procurement process is embarked upon.

The Private Finance Panel and the Treasury have issued a *Step-by-Step Guide to the PFI Procurement Process* which consolidates the process into a set of 13 basic stages. The first (and most important) of these stages we have called 'establish business need'. It is at this stage that professional advice on all the aspects of the service to be delivered, and on the asset from which that service could be delivered, needs to be brought in by the procuring team. This will help it to set the parameters in such a way that the bidders' proposals, and the one proposal which is eventually negotiated and settled upon, are likely to display all the desired characteristics and be operationally fit for purpose.

As Lord St John has said, the desired result of the design process in PFI procurement is most likely to be achieved – clearly it cannot be guaranteed – if the right advice is sought by the procurer at the very outset and if the services of the design adviser are fully and properly used throughout the whole of the procurement process. Design advice should be used by the procurer to establish design criteria; to help ensure that those criteria are part of the overall evaluation criteria; to evaluate design solutions when bidders present them; and to develop those design solutions with the bidders.

The NHS Quality Initiative

John Horam MP
Parliamentary Under-Secretary of State,
Department of Health

I have approached this issue firstly from the perspective of a Health Minister responsible for an enormous estate and secondly as someone with a lifelong personal enthusiasm for architecture and the built environment.

One of the great advantages of being a Minister of Health and sitting in the House of Commons is that you look across the Thames to St Thomas's Hospital, which is probably *the* lesson of design and architecture in the health area. To the west is Henry Currey's original 1870 building, in four sections. He had help in designing it from Florence Nightingale herself, who insisted on the pavilion layout. She wanted the possibility of isolating each block, in case of fever. Then we have the Yorke Rosenberg building on the corner of Westminster Bridge. That was finished in 1976 – a great building, but completely different from the Henry Currey building. Then there is a rather undistinguished block in between the two, backing on to the road behind. Despite the efforts of many distinguished architects over a century, therefore, what one has is a mess. Three different types of building, totally contrasting, very difficult to work in and visually very upsetting. I would almost have preferred that the Henry Currey building had been reproduced, or the Yorke Rosenberg plan adopted throughout, in the interests of greater coherence. But the unsatisfying result is fairly typical. Only one hospital building since the war has won a major award – the Wexham Park building in Slough by Powell & Moya.

The other building which people talk about a great deal in the health sector is St Mary's Hospital in the Isle of Wight, by ABK. There they have run into tremendous problems with stainless steel cladding and the NHS Trust now faces an £11 million bill to replace it.

The history of even distinguished architects getting involved with the health problem has therefore been distinctly mixed. What struck me, coming into the health area and having an interest in architecture, was that very few of the 'big name' international architects have designed buildings in the UK

health sector. I was cynically told that this was because the fees were not
high enough; health did not pay and therefore it was not interesting to 'big
name' architects. I do not know whether that is true or not, but it is true that
we do not have a lot of money to spare. I was also told that such buildings
were far too technical and gave little scope for imagination; the whole thing
was too complicated for architects, who are more used to building airports,
office blocks and so on. I do not know whether that is true or not either, but I
certainly regret the fact that the 'big name' architects make very little
contribution to the health area.

It is certainly true that the NHS is a particularly demanding client. First of
all, it has to be a workplace for a large, varied and highly specialised
workforce, ranging from specialist clinicians and tertiary services right
down to reception. The facilities are extremely complicated. There is a
wide range of services offered by most hospitals, from hi-tech surgery to
outreach services. There are also many complex statutory health
requirements. It is a high-cost investment and stringent financial
requirements invariably have to be met. The Chelsea & Westminster
hospital, for example, was commissioned at about £150 million and ended
up costing £250 million. That cost overrun – a major mistake in financial
control – starved Middlesex and the Home Counties to the north of
London of all kinds of necessary improvements for the next five years.

Then, of course, hospitals are healing places, where people do not want
to go and, when they are there, want to leave. They want to get well as
fast as possible. They are afraid and worried and the environment
should help them to overcome that fear and worry. Hospitals are also
huge fixtures in the local neighbourhood and have to be good
neighbours. The Chelsea & Westminster is surrounded by quite
expensive two-storey residential houses, and it sticks out like a factory.
Too often hospitals do resemble factories.

Medicine is going through enormous and continuous change. It is now a
very hi-tech area where, for example, day surgery is accounting for 25% or
more of all cases brought in for treatment. This means that hospitals have
to be designed to be flexible. They now need fewer beds and facilities. All
the conceptions about what a hospital is are changing dramatically about
every 20 years. I suspect that will continue. We therefore have to look at
designing 'envelopes' which can accommodate rapid change.

The danger with the Private Finance Initiative is that it will make this
process more difficult. First of all, PFI is undoubtedly far more complex a
procurement process than architects working for the NHS have been used
to. Secondly, the way the architect/designer is employed and involved can
too easily lead to the faults of the design-and-build approach being
repeated.

In addition, we are talking about a huge volume of work. Over the last 10
years we have had the biggest hospital building programme in the history

of the NHS: £2bn a year. But we have still not kept pace with the demand for new hospital facilities. Despite that expenditure, we have only managed to build one new hospital a year during that period. The way PFI is going at the moment suggests that it will probably achieve four to five totally new hospitals a year. The ones at the head of the queue at the moment are in Norwich, Dartford, Calderdale, South Durham and Carlisle. Those four or five will certainly come to fruition in 1997. We are therefore talking about a huge investment in design, with an enormous impact on the towns and cities where they are situated.

This is a huge challenge to the Health Service, to the architecture and design side of it, and to the trusts involved. My first reaction was to set up an advisory committee on design and quality in the NHS, which started in autumn 1996. It includes what I hope is a sensible balance: an architect, a nurse, a patient representative, a designer, a trust chairman, a project manager and a constructor. That group is linked with the Department of National Heritage/Private Finance Panel steering group, which relates to PFI as a whole. I have had talks with Richard Inglewood at the DNH about cooperation between the two. I hope we do not have too many committees here, but I think it makes sense to have one which relates specifically to this huge investment in the Health Service. I am sure the two groups will work well together; I certainly have had the benefit of Richard Inglewood's advice already on how we can cooperate and help each other to achieve our respective goals.

NHS Estates have also set up a working group to report to and be responsible to the overall Department of Health advisory committee. Over the years NHS Estates have produced many excellent brochures and pamphlets on design, notably *Better by Design*. They are now working on an even more comprehensive document, which takes into account the problems and lessons of the Private Finance Initiative.

I am about to commission some research into evidence-based architecture. I have already mentioned the need to create a healing atmosphere. I have asked Professor Keith Critchlow to undertake some research on that aspect, to ensure that the UK is well advanced in evidence-based architecture. America, which on the whole has the lead in this area, has made significant strides over the last few years.

We want to reward excellence. From 1997, the Department of Health will be sponsoring the healthcare section of the annual RIBA awards. Owen Luder and I have written to senior officials in the Health Service about these awards. Not all are about building big hospitals. There is a range of services needed – from small clinics and GP practices to large hospitals. Stephen Hodder, who won the RIBA's Stirling Prize in 1996, also won an award for a GP's surgery in Manchester. Help and encouragement for smaller projects also has a place.

Let me mention in a little more detail some of our early findings on the problems arising from the PFI. Early indications are that some of the problems are as follows:

- First, with a few honourable exceptions, little attempt is made to specify design quality in anything but the broadest terms. From our experience so far, it seems to us that the hospital trust, which is the commissioning body, says hardly anything at the very beginning about what it wants in design terms. Or, on the other hand, it becomes far too specific and gets down to specifying what materials and types of finish it wants. A balance should be struck between the two: an overall view of what is required and some indicators – but nothing too specific.

- Secondly, many bidders are given a strong steer by the briefing process at the beginning of the contract to provide a replica of what has always been envisaged. Very often you will find that a new hospital has a long history. A new hospital will have been wanted in a particular town or city for 30 years, and all sorts of ideas will have grown up, encrusted with barnacles, around this new project. These ideas will almost have formed it into a shape, which it is very difficult for those coming from outside to change. The project has a momentum of its own, very often arising from the views of the clinicians. It is very difficult for the bidders to bring in new ideas. The steer they get is a very cautious one, reflecting the history of the project.

- Thirdly, trust teams may struggle to evaluate solutions significantly different from the one they have already thought of themselves. In other words, not only does the bidder have a problem getting away from the history of the project, but the trust has a problem evaluating the sorts of solutions that a bidder might come forward with. This may be because it lacks expertise. It may not have anyone with architectural skills on the staff. As a result, the trust will not be able to judge the quality of the alternatives which the bidders are putting forward.

- Fourthly, little attempt seems to be made by clients or providers to use evidence-based ideas linking design to health outcomes. I mentioned earlier that this is where America is well ahead of the UK. So far we have seen very little attempt on the part of trusts to go beyond functional, financial and statutory requirements and look at things from the point of view of health outcomes.

- Fifthly, time and cost pressures in the bidding phase affect the design process. I have been very aware, as the minister responsible over the last two years, of the impending General Election and the uncertainty that causes. The present Government is committed to PFI, and so bidders know where they stand. There has been slightly less certainty about what would happen after the General Election if Labour were to win. People have therefore rushed forward to get schemes out

before the election. What has suffered has been design and the depth
of thought necessary to achieve a good design.

- Finally, trusts have tended to focus on solutions rather than on the
  quality and composition of the project team. Trusts, by their nature,
  are not always very skilful at project analysis. They have tended not
  to look closely enough at the architecture/design team or at their
  private sector partners.

There is also a tendency, partly resulting from lack of experience in project
management, to allow costs to overrun. Changes in specifications lead to
affordability problems and the project then has to be scaled down again.
All sorts of solutions are ruled out at a late stage. If the original design
concept had been stuck to, a better design would have resulted.

There are two other points which are also very important. There is no
doubt that financial and legal issues can dominate the design process.
They can be so difficult to resolve and so time-consuming that they can
shut out design problems. In a typical PFI hospital scheme, you first of all
get approval of the outline business case, then the full business case, then
the commercial deal between the consortium and the trust. Design issues
can simply get lost in the procurement process.

Let me finally say something about how we can proceed from here.
Following the first two meetings of the advisory committee and the NHS
Estates working group, we identified five issues which we feel encapsulate
the way forward.

- communication
- project organisation
- evaluation
- research
- consultation

Incidentally, I think we might have included aesthetics and design at some
point in that list.

Communication

Communication should obviously be concise and unambiguous and, in
particular, timely. The information requirements at different stages of the
design process are quite different. One needs to have a critical path
through it, with the information you need slotted in carefully. That has
certainly not been done on the initial schemes we have had so far. It has all
been far too confused.

Project organisation

Project organisation requires a good project director. That is a crucial appointment. The right person may not exist within the trust organisation. Someone may have to be hired from outside who will then have to learn very rapidly about the long history which a particular hospital project may have, and understand the sensibilities and personalities involved.

Equally, if someone is appointed from inside, then his skills must balance those of the rest of the team. All kinds of people can become project directors but it is important to get someone who has the confidence of the entire team. Also important is his relationship with the chief executive and chairman of the trust. He must have the backing of the trust board. Too often there has been almost no link between the board, which is ultimately responsible for the new hospital, and the project director, who may be too low-level a figure.

We advise trusts to think about bringing in architectural advice at a very early stage, to allow someone with professional experience to keep an eye on where the project is heading. That is especially important if the trust does not have sufficient architectural expertise on the inside. The competitive interview is becoming a more important method of choosing between design teams. We would encourage that in the health sector, because trusts need to choose carefully the people they are working with even before they reach the stage of thinking in detail about design solutions.

I think it is good to have someone who is enthusiastic about design and architecture on the board of a trust. The people in the decision-making process are crucial. Particular attention should be paid to the appointment of non-executive directors of hospital trusts, to see that we have someone on each trust who champions and understands the importance of design, who has some skill in this area, and who makes a great fuss about it when nobody else does.

Evaluation

There should also be post-evaluation of projects, so that lessons are learnt. This is a particularly weak point in many architecture and design projects. Out of 34 major healthcare schemes in recent years, only two have had a real post-project evaluation. This is best done by an outside body, which can look at the lessons of the project and help them to inform future healthcare design.

Research

Some valuable research has been done in America on the effect of the built environment on patients. Let me quote from an article by Roger Ulrich, associate dean of research and professor at the College of Architecture, Texas University.

> 'Acutely stressed patients lying on gurneys (trolleys), who are exposed to serene pictures primarily displaying water...have lower blood pressure than patients exposed to...exciting pictures or to no pictures. Sensory deprivation stemming from a lack of windows on intensive care units is associated with high levels of anxiety and depression and with high rates of delirium and temporary psychosis.
>
> 'These results from scientific studies form part of the mounting evidence that certain design choices or strategies can work for or against the well-being of patients.
>
> 'Research has linked poor design to anxiety, delirium, elevated blood pressure, increased need for pain medication and longer hospital stays following surgery. Conversely, research has shown that good design can reduce stress and anxiety, lower blood pressure, improve post-operative recovery, reduce the need for pain medication and shorten hospital stays.'

I am sure that is true, and that is why I want to ensure that we do some positive research in the UK which can bring this home to people in a significant way.

Consultation

Finally, on programme and resource issues, an important problem facing the project director and the architecture and design team is to consult the many diverse stakeholders. As I said at the beginning, a hospital is a complex organisation. You may have 36 clinical teams, all with different views about what they want. How do you get 36 clinical teams, with 100 different people, all highly educated and motivated, to form a view with you, as a designer, about what are the right solutions? There are huge problems of man-management there which are difficult to solve.

I started by saying that design issues have been a personal enthusiasm of mine for many years, and I am delighted that we are now starting on a huge hospital building programme through the Private Finance Initiative. This programme is only possible by supplementing public capital with private capital. But I am well aware of the dangers of all of this, and it is for that reason that we have taken steps to put design quality firmly on the agenda.

The Ambulatory Care and Diagnostic Centre, Central Middlesex Hospital

John Allan
Avanti Architects

I would like to do three things in this presentation. First I will outline the clinical purposes underpinning an ambulatory care and diagnostic (ACAD) centre. Then I will describe the manner in which the Central Middlesex Hospital (CMH) has handled the design and PFI processes. Finally, I will try to extract some conclusions that may be relevant to today's discussion regarding design and quality.

CMH is a 450-bed district general hospital in Park Royal, West London. It has a combination of typical problems and, arguably, untypical assets. The main problems are the hotch-potch of poorly related departments in low-quality building stock, suffering from a long-term backlog of maintenance and a shortage of capital investment. Some units at CMH are still working within the original 19th century Willesden workhouse. It also faces competition from surrounding hospitals, including St Mary's Paddington, Hammersmith, Ealing and Northwick Park.

On the positive side, CMH is organisationally sound and has pioneered advances in minimal access surgery, interventional radiology and patient-focused care. It is well integrated into its local patient and GP community and has good links with neighbouring clinical outreach facilities.

The hospital also owned significant tracts of surplus land, the value of which has been greatly enhanced by the Brent Urban Regeneration Programme and consequent inward investment. CMH was a first-wave NHS trust but, unlike many others which sought to reduce their asset base in order to lower their capital charge liability, it hung on to its outlying allotments until they could be disposed of at nearly twice their original value. Some 35 acres will eventually have been sold at prices of up to £600,000 per acre. Crucially for the ACAD project, the hospital trust has been able to establish a development fund from the proceeds of these land sales.

The ACAD Centre in model form, showing the 'service wall' (enclosing plant) and the curved 'town wall' (enclosing offices and a restaurant). Avanti Architects

Faced with these difficulties and opportunities, CMH identified three main options: to do nothing and risk going out of business in the face of increasing trust competition; to redevelop the hospital totally, which would entail massive capital investment and disruption; or to redevelop the hospital partially, by a strategic programme of new building, rationalisation and refurbishment.

The distinctive feature of CMH's decision to pursue the third option is that of applying capital investment not just to renew a major element of their building stock but to re-invent the operational purposes for which it will be used. The resulting ACAD project is thus to provide a large, purpose-built, elective care centre and will be the decisive first step in a phased process of changing an old Victorian infirmary into a modern inner-city facility.

The principal features and intentions of the ACAD Centre are:

- the separation of elective from emergency care

- the provision of improved-quality elective care at significantly lower
  case costs – a 25% reduction on current in-patient cost
- the combination of surgical, medical and imaging facilities within
  new, purpose-designed accommodation
- the complete reorganisation of clinical care protocols and staffing
  arrangements to maximise the range of outpatient care
- the achievement of added value through commercial partner
  participation in such areas as advanced interventionary and
  diagnostic imaging equipment, electronic patient records and
  scheduling systems, and integrated facilities management.

The ACAD concept originates in the drive to increase the use of day-care
procedures in the NHS and the corresponding reduction in acute hospital
beds. The aim is to disentangle secondary-level elective day care from the
unnecessarily expensive and inappropriate context of in-patient services.
In other words, the ACAD is in effect a hospital without beds. The all-in
project value will be roughly £18.5m, including equipment and fees.

Following from this strategy to use capital as a catalyst for change, it
became apparent to the trust that the architectural task would not be
one of adapting existing design templates, based as they are on
conventional departmental structures, but one of evolving new
physical solutions to suit new clinical protocols. The project must link
the building design with the process design. In other words, while the
trust was certainly committed to the incorporation of private finance
partnership, it was determined that this should not entail losing control
of the radical conceptual basis of the project.

How did it go about achieving this? It initiated a design competition,
having short-listed five design teams from 45 respondents to an *Official
Journal of the European Community (OJEC)* advertisement for expressions
of interest. This competition was entered at risk by the five participants,
who were briefed collectively by the hospital project team in a
constructive and comradely atmosphere. The contest was limited to a
five-week, single-stage submission of a defined number of A1 panels and
a competitive presentation to the trust board. The competition decision
was reached using the 'two-envelope system' – ie on design merit, with
the fee proposal being scrutinised afterwards. The organisers proceed to
the second placed entry only if the preferred team's fee cannot be agreed
outright or by negotiation. So the client wanted to make their choice of
design partner on a strictly qualitative basis, unconfused with and
undistorted by the commercial considerations which are a necessary part
of the main PFI package.

The primary objectives were thus clear to all concerned. The trust was
seeking:

- a design team which could demonstrate its understanding of the
  radical clinical proposition that ACAD represented

- a design team which it felt would add value in the widest sense, of not only a quality therapeutic environment but also a strategically intelligent building that could respond to longer-term changes and adaptive re-use as clinical programmes evolved
- a design team with which it felt it would enjoy working.

The briefing documents issued for the design competition included the outline business case, a clinical protocol and various operational diagrams, produced by NHS Estates; but the trust made it quite clear that these materials should be treated not as 'the brief' but as a virtual brief or, as the project adviser called it, an exam question to which we were invited to give a specimen answer.

By recognising the artificiality of a design competition, the client was able to reassure competitors firstly that the incompleteness of briefing information would not prejudice the validity of the entries and, secondly, that the eventual winner could expect his entry to be challenged and changed when the real process of briefing ensued after the competition. You will have gathered that this competition was won by Avanti Architects, working with the consulting engineers Ove Arup & Partners, and from then on we have been appointed and paid on a proper fee agreement.

Between May and Christmas 1995 we worked intensively with the trust's project team to develop and modify our competition entry, in what was understood by all parties as a process of interrogation and iteration, exploring numerous alternative designs to test and extend the flexibility and resilience of the main ideas.

The basic constituent parts of the design were the hi-tech imaging intervention space; the patient zone; a series of green courts or gardens; a public zone connecting the two; and the town edge part of the complex, which contains offices, a restaurant and various other facilities.

Just to underline the client's concern for architectural quality I should add that, before we embarked on the detailed design phase, the trust sent me and one of my partners to the USA for a two-week study trip, looking at innovative polyclinics and ambulatory care centres in Florida, New York and New Hampshire.

In September 1995 the trust embarked on the next stage of inviting PFI involvement. Again, they went about setting up the process extremely intelligently, by first clarifying what sort of private partnership was preferable. The OJEC advertisement did not specifically exclude any one kind of PFI partnership, but the trust was determined to identify and seek out the type of partnership that was most appropriate to this project. They were fairly sure that this was unlikely to include building procurement – in any case, the trust expected to be able to finance this from its own land sales.

Potential PFI partners were told clearly that the work we had carried out should be regarded not as 'the design' but as the design example; it was available for them to use if they wished, or else it could be used as the basis of a design developed by a team of their choosing. Now that the trust had evolved its own preferred solution for comparative evaluation, it did not want to exclude a PFI partner bringing its own design capability to the project. In the event, none of the short-listed partners did.

In addition to this, the trust defined its selection criteria for the PFI bids as follows:

- the company or consortium must have sufficient economic mass for the scale of the capital investment required
- the company must have an established commercial position, with corporate ambitions and a core business relevant to this project
- the company, whether of UK or European origin, must already be established in the UK, so that they would not be using the ACAD centre as a learning project
- the company must be comfortable with innovation and be prepared to contribute an element of R&D investment. In other words, it must be capable of bringing not only solid finance to the project but also intellectual added value. The trust's project adviser pointed out that in his view the construction industry generally brought neither. The supply of major equipment and software was identified as the preferred PFI contribution. One scenario was that the hi-tech imaging and interventionary centre would serve as an operational showroom for the major equipment supplier.

The chosen team comprises Honeywell, Hewlett Packard, HBOC and Philips Medical. The equipment is expected to include magnetic resonance imagers, CT helical scanners, ultrasound machines, fluoroscopy and a full range of theatre and endoscopy equipment, a range of software solutions for paper-free patient records, and scheduling and communications.

The design team submitted a full planning application and were then stood down while detailed negotiations continued between the trust and the preferred PFI partner. The full business case was approved on 30 May 1996 and announced by Mr John Horam on 3 June 1996.

The pre-contract phase was completed in February 1997 and the building project is currently being tendered on full bills of quantities for a traditional GC Works-I procurement. Assuming a satisfactory tender is received and the trust is able to let an acceptable contract, works are scheduled to start on site in May 1997. The ACAD should open for business in 1999.

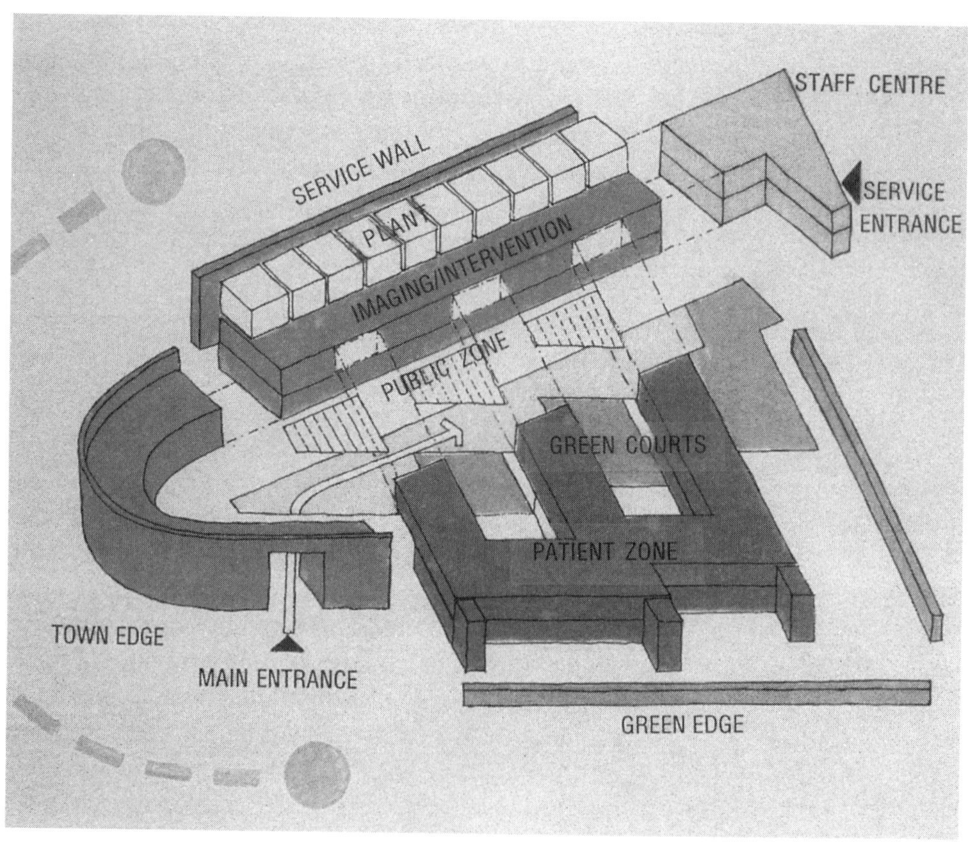

Functional analysis of the ACAD Centre. Avanti Architects

What are the key elements of this experience that might usefully inform our discussion today? Firstly, that the CMH trust's ambitions were larger and more radical than simply the renewal of its physical estate. They originated in the qualitative revolution in clinical and hospital philosophy and practice that will sooner or later engulf us all. This saved the trust from the conventional folly of approaching the PFI as if it was merely a variant of design-and-build. They realised that to stay on the pathway of innovation they must initiate and retain control of the design process.

As for the PFI, the procurer's vital insight, in my opinion, was to apply its root principle – the factoring in of private sector resources to the NHS budget – and invent a *modus operandi* to suit its own project, rather than relying on second-hand assumptions as to how it had to be done. Looked at in this broader light, the trust's establishment of its development fund from land sales is arguably just as valid an application of the PFI principle as any other; but the important rider is that it used the early receipts from this to invest in the design process and thereby retained control of the project aim.

The CMH experience has been called a hybrid procedure, as compared with the so-called 'pure' PFI process. Bearing in mind the root principle mentioned earlier, I am not sure that those adjectives would not be more correct if applied the other way round. Anyway, Avanti has also been involved in a few of the more conventional PFI projects. How do they compare?

In the conventional system the design disciplines are controlled by the PFI bidder, with the result that

- the client–designer dialogue is more indirect
- the real briefing process occurs later in the project development cycle, probably after some significant assumptions have become settled on both sides
- the designer's real client, the service provider, is not his paymaster – which means that his proposals are initially, and ultimately, a commercial function of the bid package rather than an operational function of the clinical agenda.

Just to point up the difference in terms of its impact on the client-designer dialogue – which any architect knows is the lifeblood of a quality project – in one such conventional PFI project in which our team has now been selected as preferred bidder, Avanti has achieved four client meetings in just over a year. At CMH, during the vital formative design briefing stage, we had 82 meetings in four months. In the former project, our briefing has been mediated through estates officers and project managers. At CMH, we have been dealing direct with surgeons, radiologists, nurses, doctors, health service planners and anaesthetists, to name but a few of the participants on the procurer's side.

Added to this in the conventional model is the likelihood that until the PFI contract is actually signed, the design team will be working at risk – especially if, as we are told, no public sector funding of the bid preparation is to be made available. The developer will insist that until he has an agreement the team members must all carry their own risk. Yet in the designer's case, arguably more than most others, his input is predominantly front-loaded and the gearing of his risk to his reward is virtually nil. Eventually, to limit this risk, he will tend to limit his input, with the result that his proposals are likely to be based on under-investigated assumptions.

As conscientious professionals become increasingly uncomfortable with this mode of service, I think it can be predicted that they will gradually withdraw from PFI bidding and leave the field either to gamblers, who are prepared to take the risk, or to plodders, who will try to limit it with standardised or conventional solutions. In neither case is the outcome likely to be conducive to design quality.

There is much talk these days about 'value': value for money; added value; value engineering etc. But what does this mean in the realm of architecture and design? How and when does an architect add value? If we can agree that this might have something to do with design quality – that obscure object of desire – how might it be accomplished? My answer would be that the architect adds value not just through specifying the right cladding or perfecting the proportions of his windows, important as these might be eventually, but through his capacity for lateral, speculative and strategic thought; through his tendency not to take a problem as given but to take it apart and construct alternative formulations; through his skill in cross-examining a brief; through his ability to transpose social needs into spatial sets, to organise and classify human activities in terms of what is needed to sustain them in building form.

In my view, the earlier and more directly the architect is allowed to make his contribution to a project the more value he will be able to add. Conversely, the later he enters the process and the more the raw material of his problem is transmitted and predigested by intermediaries, the less scope there is for his real professional skills to make their full qualitative impact.

How may this thought be applied to the PFI? At present we are fixated by the £500m worth of PFI deals awaiting the go-ahead, but I can imagine an even worse problem: that further down the line we may discover that much PFI investment has been ploughed into the wrong solutions or, to be precise, into accommodation which only rehouses existing institutional structures and working patterns, rather than providing solutions as flexible and protean as the changing technology of healthcare and rising social expectations are going to require.

It seems to me that unless this huge wave of capital is used as a catalyst for change, as an opportunity to re-engineer healthcare processes (as is being attempted at CMH) there is little hope of the NHS ever catching up with the apparently insatiable demand of its consumers. I am certainly not saying that architects have the answer to this historic challenge. No-one believes that. There are too many bad architects and not enough good clients. All I am suggesting is that where real, inventive design skill is brought into early, intimate and constructive dialogue with radical and committed clients, there is a better chance of high-quality solutions emerging. Experience surely shows us that it is where the architectural agenda parts company with the client's agenda that things start to go wrong.

The well-worn PFI mantra of 'transfer of risk' therefore seems to me to contain a risk of its own. Namely, that within that transfer goes the transfer of responsibility, which in turn contains an element of transfer of control and transfer of choice. If PFI assumes that responsibility for core business stays with the procurer, then it seems to me that the direct interaction of that core business expertise with the design disciplines is something too important to transfer to the commercial bidder – at least in cases where a radical solution is being sought. In any case, there are other, more suitable ways of making a competitive choice of architect than through the mechanism of the consortium bid.

To sum up, design quality may be elusive but it does not come from nowhere. In PFI schemes, as in any other sort, it is most likely to be achieved when committed clients make an appropriate investment in good designers at the right stage.

PFI and Design in the Public Realm: the Labour Party View

Mark Fisher MP
Shadow Arts Minister

Thank you for this opportunity to be provoked into thinking. I was very keen to accept the invitation, notwithstanding the Prime Minister's decision to call the election, because it is not often one has a room full of such expertise – and we have already had some very interesting and thought-provoking contributions this morning. John Allan's paper clarified a lot of the thinking and its quality was quite exceptional.

In a way, to consider how we take PFI forward brings us to the nub of concerns about design and the architectural quality of the public realm. Lord St John said it very well at the beginning, when he said that over the years the Commission had been concerned to establish that design quality *is* value for money. I am sure everyone in the room would agree with that. But what that means in terms of the trade-off between financial imperatives and design quality in any building programme is at the nub of PFI.

As Deborah Bartlett's paper reminded us, there is a paradox at the centre of PFI. It was undoubtedly initiated by Norman Lamont at the Treasury in 1992 for financial reasons: to get better value for money and to get private money into the public procurement process without increasing the PSBR. Those were the driving priorities. The essential concern was not how to get the best buildings in the public estate. Those financial imperatives and necessities will not change with a new Government.

When PFI was first initiated, the Opposition was very sceptical and quite critical. As we have come closer to Government, we have had to think more carefully about procurement. We have become more practical and pragmatic in our approach, and we will not abandon PFI. What we have heard from John Allan, however, is that, whether or not we continue to have PFI, where and how it is employed in the public estate is crucial. We need to improve our thinking about how to get good design, buildings that work, at low lifetime cost, fulfilling the purpose for which they were designed, within financial and other parameters. Because the core of the

PFI was financial, there is a tension between the desire solely to get good value for money and the desire to produce good buildings.

In thinking about how a Labour Government might address this, it is important not to become too hypnotised by the technicalities of the PFI. We must step back from it and see it not as a master but as a servant of what we want to achieve in the public realm. We must recognise that, first and foremost, Government needs to improve its performance as a procurer of buildings, as a client of buildings and as a setter of the best practice in the enormous public estate. So rather than starting by asking how we make PFI work, we should start by asking what we are trying to achieve overall; we then need to look at where PFI might fit into that and where it might need to be adapted and reformed.

Taking that pragmatic view, it is important for us to see what a Labour Government would be inheriting. It is a mixed inheritance, some of it very good, some of it with distinct weaknesses. One of the best parts in this particular policy area is John Horam's initiative in setting up his advisory panel. That was enormously intelligent and optimistic, and it was encouraging to hear a Government minister care so much about it and pay so much attention to detail. Whoever goes into the Health Department will be enormously encouraged to inherit that panel of advisers with that depth of thinking. The same is true of the Department of the Environment. It is perhaps a rather strange thing, in the middle of a political battle for who should govern the country over the next few years, to be paying compliments to your opponents. But it is undoubtedly true that John Gummer, from the point of view of architecture and design, has been the best Secretary of State for the Environment for years. Similarly with the Department of National Heritage, where there are good things to inherit from Virginia Bottomley and Iain Sproat. Certainly publications like *Better by Design* and the work of the Arts Council in recent years in thinking about procurement all form part of a good inheritance.

Another useful inheritance is that we are coming out of recession. Most particularly, we are beginning to see – which really was not the case seven or eight years ago – a whole range of first-class architects designing in the public realm. I recall that at the time of the 1992 election the country was not using its design talent. That is no longer true – partly because of the Lottery and partly because of the change of attitude in Government encouraged by John Gummer, John Horam and others. We are beginning to see good-quality architects employed in the public rather than purely in the private realm.

All this can be built on. But there are weaknesses too. We will inherit a real problem in local authorities, where there is a lack of morale, a lack of confidence, a lack of capital, a lack of experience in building projects; but also a huge pent-up desire to build. Although a Labour Government would very gradually release the constraints on local authorities, it is nonetheless true that housing committees, planning committees and others are straining at the leash. They wish, perfectly understandably, to respond to the huge demands both for large-scale public buildings and for public housing. The pressure is to hurry on, to get a building programme started.

There are many problems there. Most particularly, in spite of the good practice I have mentioned, there is no coherence across Government. I asked John Horam earlier today what relationship his design panel had had with John Gummer's architectural adviser. He said 'None at all'. They do not talk to each other. That is unfortunate. He said that he was making links with the team at the Department of National Heritage, but this lack of coherence across Government is not good enough. If Government is to improve the whole public estate, it has to look at it across Government, to think corporately rather than sectorally, horizontally rather than vertically. That is the most difficult thing for Government to do. It is against all the traditions and all the budget structures of Whitehall. There are substantial personal, political and financial reasons why Government does operate vertically – but we will not get the best from an idea like PFI, which is meant to speed up and improve public procurement, unless Government starts thinking horizontally. That is the biggest change that needs to be made.

Given this mixed inheritance, our aim, piously but I think sensibly, will be to improve Government performance as a whole. The value of PFI projects is still small compared to the total Government procurement in repairs, renovations and new building. That amounts to £5bn a year, leaving aside Lottery funding. There is a huge amount of public work going on and yet no coherent attitude. The potential to improve performance is enormous.

We will only succeed in making improvements if we put design alongside value for money. Of course it is proper that politicians, spending taxpayers' money, should give a lot of consideration to value for money, but design should be alongside it – particularly because we have to start thinking about design as a way of getting value for money. Design is not an additional extra cost; design is a way of thinking. What I most took from John Allan's words was the quality of intellectual engagement with a project. Imagination and thinking are what architects and designers most bring to it; the ability to solve problems. Aesthetics arise out of that. Fitness for use and design quality both arise from a high quality of thinking and problem-solving – so that if Government can improve its performance as a client, you will have a better client thinking more intelligently and a better system of selecting designers and design briefs. That improved quality of thinking could lead to reduced costs.

Until we start thinking of buildings in terms of their lifetime costs rather than in terms of their initial capital costs, we will not be paying sufficient regard to sustainability. Lower maintenance costs, lower energy consumption and less pollution from buildings are not reflected in the initial capital costs of a building. At the moment the Treasury is committed to first cost – the capital cost of getting the building up. No consideration is given to the added value and cost benefit analysis. But without some degree of cost benefit analysis, you cannot consider the lifetime costs of a building. It is no good getting better value for money on the physical bricks and mortar of a building if the long-term costs are disastrous.

All of us, in our communities, know that only too well. My city is Stoke-on-Trent. We rushed up buildings after the war and were glad to get them. But they have cost us over and over again, because the building materials and construction techniques were poor and the maintenance cost of the post-war public housing stock has been horrendous. The capital cost was insignificant, but there has been a cost to that city in maintaining inadequately thought-out and inadequately designed and constructed buildings.

We are very foolish as a society and as a Government if we do not start thinking in terms of the consequential costs of buildings as well as the capital costs. If we say that thinking horizontally across Government requires totally new thinking, then to get the Treasury to consider the consequences of its expenditure rather than the immediate one-year cost of expenditure will be a revolution. It will be some time before a Labour Chancellor would be in a position to start thinking in those terms; but without that, everything we say about getting the best value from PFI will be at risk of being undermined by the long-term costs of the building.

Aside from the way the finances of a building are evaluated, we need to consider the quality of the procurer. It was encouraging to hear John Horam say that we needed to get people who really care about design into public positions. I suspect considerations like low maintenance costs and lower energy costs would then be built into the design as part of the design thinking. It needs someone to encourage, empower and require architects to put that into the design process.

Our aims, therefore, are to improve our performance, to consider the relationship between good design and lifetime costs and to be a better client. Before the 1992 election, we were seduced by the idea that the selection process was crucial; we had to be more intelligent about selecting designers. But we put far too much weight on good selection, on competition policy, without thinking enough about what competition means and what its different varieties can offer. It was interesting to hear from John Allan about his experience of the 'two envelope' approach. That is something that ought to be adopted more widely. It is one of the ways to give primacy to design rather than finance.

Over the last two or three years, helped by architects, we have tried to run small seminars to see how, if we were in Government, we could be a more intelligent, better client, anticipating problems, thinking of better design briefs, thinking of what we were trying to get out of the design process. We would hope, in Government, to develop guidelines that would inform the procurement work of all Government departments – guidelines on timing, briefs, selection and form of competition. To achieve this we will appoint a Government panel of architectural advisers – a Government-wide version of what John Horam has done at the Department of Health, though with greater representation for architects – to help advise all departments. It will not tell them what to do, because the autonomy of different departments will be jealously guarded; but it will help them get better value out of new public buildings by procuring buildings which are more popular and better to work in while serving the purpose for which they are designed.

Before I bring my remarks to a close, I will address specifically how we see the future of PFI. This part of my speech has been overtaken by John Allan's, because he gave a wonderfully specific and practical example of how best to ensure that the design process is not dominated by the financial processes. Government can learn from experiences like his and from his comments about the crucial element of timing, so that the designer does not come in under the wing of the PFI bidder but takes a full part in the process. We also have to be strategically more selective about when we use PFI. That may make for some difficult choices.

We also have to consider carefully where and when to use PFI. It is an excellent system for developing countries where money is short, to get private money into the public estate on suitable occasions; but we have to look carefully at whether it works in the UK for facilities like prisons. It is difficult to see where there is added value for a private financier in building prisons and perhaps in building courts as well. Much of the public estate may be suitable for PFI, and we will continue with it in those areas; but there are undoubtedly areas where it may not be suitable – particularly when we are not in a recession, when the attractions of PFI for the private sector must be considerably less. There are some indications already that the enthusiasm of the private sector for PFI is not increasing. In a period of growth we need to be selective about what projects are suitable for PFI and then, more particularly, to learn from the excellent example of people like John Allan in working out how to get the design process into a central position within PFI.

Discussion

Richard Burton (Ahrends Burton Koralek): Speaking as the chairman of the Architectural Advisory group to the Arts Council, I am very encouraged by what I have heard today. I would compliment the Royal Fine Art Commission on having this seminar at all at this very important moment. In a way, the timing is perfect. We are all aware of what the problem is, and it is for the next Government to act to find solutions. We have heard today a Health Minister who is obviously very excited by architecture but could only achieve a limited amount. One hopes that the next minister will be able to do a good deal more.

What worries me about the idea of having a designer or an architect on the board of NHS trusts is that I have seen this happen. We have been closely involved with an architect advising a trust, and I have seen how the PFI process as it currently stands can bulldoze design out of the equation. Cost was all.

I am encouraged that the RFAC has recently become a non-statutory consultee in the planning process. It will make an enormous difference when the trusts and bidders realise that they have to go through a very stringent planning process.

The position at the moment is very unsatisfactory, and I think the Treasury is the problem. They have been very short-sighted. I would like to suggest that they look at the assessment process which we use in the Lottery. It is much more comprehensive than the assessment processes which have been applied to PFI and manages to take account of quality of design. English Partnerships too has a very good approach to design. The Treasury has to move towards having a completely different attitude to design which offers not only value for money but money for value.

Edward Cullinan (RFAC/Edward Cullinan Architects): It was immensely sad to hear John Horam remind us that no health building had won any kind of award for years. Think of the early 18th century and the courtyard at Bart's designed by James Gibbs. Think of Duiker's Sanatorium in Holland in the 1920s, or Aalto in Finland, or the Finsbury Health Centre by Lubetkin in the 1930s. They were not just good buildings; they were forerunners of any kind of building. Out of all building types they were the most inventive, the most creative and the most energetic. I suspect they were all designed for particular clients with very particular, very definite, very practical views and ideas. The first crucial thing we need to do is to bring the users – that is, the surgeons and doctors – right into the middle of the design process.

My second point concerns communication. In a way I am repeating John Allan, because he made such an excellent presentation. I want very crudely to suggest that in complex buildings such as hospitals the design process has two stages. In the first stage you come to an understanding of the circulation diagram: the human connections, the sizes of things, which pieces have to be near each other and the systems and technology of the building.

At the Royal Fine Art Commission we usually see hospitals which have only been through that stage, where the circulation diagram has somehow had a roof put on to it and some walls added. It leaves out altogether what I would call the vital second stage. That is, having grasped all those complicated issues, you need to decide to compose the thing in a manner that you and others desire, so that it is comfortable to use, efficient and, if it is a hospital, soft, gentle, homely, daylit and sunny. The building has to have a deliberate, composed style and point of departure that somehow encompasses and rides far above and beyond the complex issues of its circulation diagram. Lubetkin achieved that at the Finsbury Health Centre; so did James Gibbs at Bart's. The architecture far surpasses the difficulties of a circulation diagram.

Out of John Horam's list I would pick only one: communication. That is critical. It should be concise, unambiguous and timely and, I would say, inspired.

John Wells-Thorpe (South Downs Health NHS Trust): There is an essential link between user and designer, which has been underlined by speakers this morning. In one sense I am lucky enough to be both, as chairman of an NHS trust and as an architect.

One of the things which has struck me over the last five or six years in our own capital programme is how willing and anxious the clinicians and consultants are to achieve good design, and how enthusiastic they become over the dialogue which can take place. This particularly struck me 18 months ago when the Royal College of Physicians, in their magnificent building designed by Denys Lasdun in the early 1960s, put on a celebratory evening in honour of his 80th birthday. He was invited back to talk about that building. He drew the distinction between the enclosure of space and the definition of space. Had I been asked beforehand how that would be received by surgeons and physicians I would have been tempted to say that it would go right over their heads and they would be itching to get the next train back to Charing Cross. Instead, they were utterly fascinated. They said it was the first time that they had been involved in a capital investment of that sort, where they understood what the design component meant and what the architectural contribution could provide.

Far more recently – because my principal concern is mental health – I have found that groups of psychiatrists and psychologists with whom I am working are fascinated with the notion that a building for, let us say, people with dementia can be designed around the concept of what we call 'cognitive mapping' – that is, being able to touch those senses which are not impaired by dementia. As you well understand, not all of your five senses diminish at the same rate if you are suffering from that sort of condition. Texture, smell, colour, light and sound are all things which are capable of being manipulated by the designer for the benefit of the patient. Once you have the clinicians and consultants saying that a particular working environment is better for patient recovery, then you really are beginning to achieve things.

I do not think we have touched on this question of bringing together the user and the designer at the appropriate stage. If you look at the PFI as it was originally conceived, it almost guaranteed the alienation of the user and the designer. I am sure it was unconscious, but there was a built-in perversity which practically ensured that a proper coming-together did not happen.

Stephen Barter (Richard Ellis): At Richard Ellis we spend as much time acting for Government in evaluating PFI bids as we do acting for bidders. I want to make two points. First, I was very pleased to hear Mark Fisher's commitment to PFI and particularly his comments about the tension between financial considerations and design quality. However, I did not quite recognise the point he made about the evaluation process not taking into account the running costs of the building.

In my experience there is just as much attention focused on future running costs; the facilities management arrangements and their revenue implications are as much a feature of these appraisals as is the initial capital cost. Indeed, the PFI process is designed to make explicit the whole lifecycle costing of building construction and management.

A lot of our time is taken up with Government office accommodation. There is an enormous opportunity here for better design because of the way in which technology has changed working practices and the way in which Government is organising itself to use its buildings differently. I am thinking particularly about the time spent in these buildings; the increasing use of out-sourced staff and part-time staff to carry out certain functions; the introduction of wireless technology and the implications that has for buildings which have previously been written off as redundant. All of these bring significant design opportunities related to how the building works and how much it costs the public.

Mark Fisher: It is a very interesting point. I suspect that this is a positive, if arguably unintended, consequence of PFI. By bringing in outside private interest which has a long-term concern in the running costs of the building, it gives a primacy to that area which Government, and indeed local authorities, tend not to have.

That long-term interest is quite difficult to explain to the Treasury, who are only concerned with this year's PSBR. There is a lot of short-termism in the Treasury, but maybe the PFI – in a possibly unintended way – might correct that imbalance slightly.

Bryan Jefferson: The projects we have heard about this morning and the tone of the meeting have tended to give emphasis to large, complicated projects; but PFI is intended to operate across the spectrum. I wonder whether anybody here has had experience of PFI proposals, either successful or abortive, on smaller-scale projects.

Norman Rose (Business Services Association): My interest is major contractors who provide services on PFI projects. Possibly the answer to your question – although we know that there have been lots of small projects – is that bidding costs are so great that the smaller project becomes wholly uneconomic. We know that in a large project bidding costs and professional fees can total £1.5m. Contractors involved in three or four projects at one time have an enormous amount of capital tied up in something which may come to nothing.

I have been heartened by what all the speakers have said this morning, particularly by the amount of common ground there has been between John Horam and Mark Fisher. Operators come in the middle, between the client and the user. Despite the original idea of PFI – ie the provision of services funded privately – we often find that it has turned into the provision of bricks and mortar. That may arise because construction costs are the largest single element. The construction industry has been at the forefront of helping buildings to go up but we tend to lose sight of the fact that it is the provision of services over 25 or 30 years which, in retrospect, will make or break PFI.

I would suggest that we ought to be looking at PFI becoming a design-operate contract, not a design-finance-build-and-operate contract. The operator has to be able to operate within the internal environment of the building for up to 30 years. If the designer chosen by the client and the operator come together, they can probably put together something which meets both parties' objectives.

The consortium goes out on the open market to raise money for the project. It gets a 'bid' from either a bank or group of banks who feel this is a project they can support, because the risk is high enough to give them a return. It could reduce the cost of the project if, within the overall consortium, the construction element was its second sub-contract – bid within the whole project as opposed to being bid before the project gets off the drawing board. Instead of having assumed that a particular construction company will come in at the beginning, therefore, you look at finance and construction separately once you have an idea of the shape and form of the project. Representing those who have this task of operation, my view is that such a system would greatly improve PFI. It would also weed out those PFI projects which are not 'PFI-able', because the operator will look at it and say that it is not practicable.

Bryan Jefferson: John Allan, you have been rightly complimented on what you have been telling us this morning. You have heard the views of others and I wonder whether, in the light of those views, there is anything you would like to develop a little further from your own experiences.

John Allan: The thrust of what I have been trying to say is that the different participants in this process should do what they do best at the relevant time. As the last speaker has pointed out, it may well be the case that the finance and building input is not necessarily most relevant at the front end of the project. In the case study I referred to, in fact, the way in which the building contract had been procured was entirely traditional.

The point I would make, however, is that architects are really terribly cheap. Their risk and their reward are in a sort of 1:1 ratio. If you cost that over the whole project cycle, it does not save any money to bring them in later. You might as well do so at the beginning, when their work has most value.

One point which emerges – and it takes us out of this discussion slightly – is whether NHS trusts are too small, as individual units, to make the best use of strategic planning in a PFI context. We come across this to some extent in the application of PFI to the education sector. Although Avanti have not achieved any projects in that sector, we have been on the brink of a number of them. It does become possible to imagine that a single school, for example, is not of a large enough size as a business unit for value to be added through a PFI application. If you take that example back to the healthcare sector, then some trusts may be able to derive maximum benefit from a PFI model if they grouped together, or if there was an opportunity for strategic thinking between trusts, rather than just the competitive relationship which trusts are obliged to adopt towards each other in the internal market.

Bryan Jefferson: PFI systems are being put into educational projects on quite a modest scale. The hope is that they are going to work. People do believe that there is merit in having PFI at that level.

John Wells-Thorpe, who is chairman of a health trust, might have a view on whether the trusts, as presently organised in the NHS, are too small. Do you feel inhibited both by the size of the operation and by the competitive environment?

John Wells-Thorpe: I have to give you a qualified answer, because trusts vary a great deal in size. Some are very large and powerful and have enormous financial clout; others are tiny. Irrespective of any change of Government, rationalisation between adjacent trusts is something which should be encouraged. When the trust boundaries were set up it was a very good first shot, but some are relatively vulnerable because of their size and purchasing power; others are exerting a lot of muscle, which is not good for their neighbours. I think these problems will be ironed out in time. I would admit that the separateness of trusts, and therefore the sovereign attitude they take over their own territory, can lead to problems of the type John Allan mentioned.

There is another built-in disadvantage in the present system. In a perfectly plausible and appropriate effort to give more decision-making power at a lower and more local level, the Government has created a larger number of smaller trusts. With the swing from secondary to primary care which the NHS is now experiencing, organisationally and philosophically that must be right. But what it makes more difficult is the question of rationalisation into larger purchasing units. I think we have to find a way round that.

David Steeds (Chief Executive, Private Finance Panel): It might be helpful at this stage if I put PFI in a macro-economic context. We reckon there will be about £2bn worth of PFI projects to fund for the 1996/7 financial year. We reckon that that figure will more than double next year, to about £4.5bn. PFI is therefore beginning to play an important part in the national economy.

In the context of Government-sponsored capital expenditure, however – £20–£25bn, depending which year you are looking at – it is still relatively small. I do not think anyone ever anticipates it accounting for more than 20% of Government-sponsored capital expenditure. There will always be a lot of conventional procurement. We must not spend all our time, therefore, looking at PFI. We must consider how best to improve design within conventional procurement as well.

That leads me to my second point. While you may criticise PFI, it seems to me there is no evidence that conventional procurement produces good results either. The PFI Executive is based in Victoria Street, very close to Marsham Street. If ever you want to remind yourself that conventional procurement can produce disastrous results, just walk past Marsham Street.

As many of you will know, I am new to the public sector – I took up my present post on 1 October. There is a lowest-cost mentality in the public sector. It is not surprising. Every civil servant is paranoid about appearing in front of the Public Accounts Committee. To explain that you have spent another 20% on a £100m hospital because you liked the design better is very difficult. In a sense the problem is to convince the general electorate that they should pay more for good design. In my view that battle has not been won.

Mark Fisher mentioned through-life costing and I was glad to hear Stephen Barter point out to him that PFI does focus on that. It is not an unintended side effect of PFI; it is one of the benefits of using PFI that bidders are required to look at the through-life cost of the whole project.

Returning to the health sector, both the PFI industry and the design industry need to find some way of working better with clinicians. That is what Avanti were focusing on. It is not satisfactory that at the moment a PFI project in health only involves the non-clinical services, because it is the success or otherwise of the clinical services that determines the reputation of that hospital and therefore has a big impact on the success or otherwise of the PFI operator. I do hope that we will find some way of involving clinical staff in the PFI process. I recognise that this may not be politically realistic, but an obvious way would be for PFI operators to contract directly with the purchaser, and then for the NHS trust to be a labour-only sub-contractor to the PFI operator. In that way the hospital doctors could stay within state employment, if that is felt important.

One last point which I would like you, as designers and architects, to think about is the lifetime of the building. Inevitably you are building for later generations to admire your wonderful works. We would all like to live in a world where we were surrounded by nice buildings. But we have to be realistic. The NHS currently assumes that a hospital lasts 60 years. This has caused a lot of problems in funding health projects, because I believe that 60 years is too long nowadays. In my view, 20 to 30 years is probably more realistic. Technology is changing very fast in health and in other areas. I suspect that in 30 years' time our grandchildren will want to make drastic changes to those buildings, and may prefer to start again. Even if they do not want to, they may be forced to pull those buildings down because technological and other advances will render them unusable. So we have to think about how long we should be building for. If we are only building for 25 or 30 years, we have to look at how to bring in good design to what could be a relatively short-life asset.

Christopher Liddle (HLM Design): I have found this morning's proceedings very encouraging and uplifting. In our firm we have been involved in PFI in healthcare since mid-1994 and had something of a healthcare track record before then. PFI has provided the most marvellous opportunity for research and development in healthcare design, which we would not have seen under any other circumstances in that timescale.

It has provided us with an opportunity to enhance and establish a research base and, above all else, it came at a time when our profession – undoubtedly the most battered of all the professions in the recession – saw an opportunity to pursue much higher objectives in healthcare design.

This cannot be done for no fee and at no cost, however. We cannot become involved in PFI schemes where we are not properly paid. We have turned down at least £150m worth of opportunities because of particular organisations which were unable to pay our fees. An architectural practice cannot rise to its design aspirations unless its key players are involved, and those people cannot work for nothing.

You cannot design a complex building like a hospital without massive interface with the users. It is an irony that so much of the discussion regarding PFI has been focused on healthcare, because it is hard to think of a more complex building relationship than that involved in putting together a major hospital, where change is so rapid that built-in obsolescence has to be considered.

In terms of the good things about PFI, we have been able to enhance our own understanding and research base in healthcare, and we have very deliberately worked only with those organisations which at the outset – not three or four months into the scheme – were prepared to try to understand the needs of our team.

John Allan: To return to the question of life-cycle costing, we already have a set of rules which can help to develop an awareness of its importance in designers and in people who procure buildings; but it is never really looked at in a positive light. This is the CDM – the construction, design and management regulations – to which we are all subject, which we are all obliged to apply to our building projects, but which tend to be regarded as primarily preoccupied with questions of safety; it seems to me they could, if developed and interpreted in a wider and more positive light, contribute to a slightly broader culture of maintenance and life-cycle costing.

Observations on Current PFI Projects

Trevor Osborne
Chairman, Trevor Osborne Property Group

I want to address some of the issues we touched on this morning and a few others besides, and to do so from the perspective not only of a Royal Fine Art Commissioner concerned about the architectural merit of PFI proposals but also as a participant in the process and as a bidder for a number of property-related PFI schemes. I will talk generally and then a little more about some of the particular schemes, and consider the fact that this new method of procurement may perhaps have its dangers as well as its initial benefits.

I want to say straightaway that I am a supporter of the Private Finance Initiative. I am painfully aware of its shortcomings and the difficulties that lie in the path of those who wish to be involved; but I believe that if we bear with it and refine it to our best purpose it will bear fruit.

The question before this seminar is what sort of fruit will it bear. It is the issue of good design which has driven the idea of the seminar, which I am delighted that the RFAC decided to hold because, as Ted Cullinan said this morning, the Commission is now seeing a number of proposals coming forward which are the result not only of Lottery funding but also of the Private Finance Initiative. Certainly they bear some similarities. The principal similarity is that often they cannot be properly judged because they have been inadequately prepared.

Why then did we need a new Private Finance Initiative in the first place? What was wrong with the way we did it before? We have tried various methods. I do not think it can be disputed that, with some exceptions, post-war provision of buildings and public services has been disappointing. The traditional procurement methods have not always generated buildings of architectural merit or indeed of lasting worth. It might also be true to say that quite often the traditional procurement methods have led to massive cost over-runs and construction delays. Society bears the cost of those mistakes in ways other than just the building and maintenance costs.

Yet we have a legacy in Britain of fine pre-war buildings. The Victorian town halls still grace our towns and are a joy to see; they have lasted. Cubitt's King's Cross and Barlow's St Pancras were created with a good deal of pride in the achievement. Looking beyond that and earlier, Sir William Chambers built London's first purpose-designed office building at Somerset House. Look how difficult it is proving to be to get the Lord Chancellor's Department and the Inland Revenue to vacate it. It has stood the test of time and is much valued by those who still use it.

What one learns from these examples is that these buildings were statements of confidence and civic pride. They were not just another building contract. They were put up by people who wanted to symbolise something of the status of the public use which drove the need for the building in the first place. Civic pride and civic responsibility drove those patrons to appoint architects who would share those ideals.

I will not bore you with the history of procurement, but in post-war Britain the Property Services Agency, itself with many fine achievements, also found itself handling some extremely difficult contracts, with a wide variety of outcomes. Local authorities' architects departments differ widely. In Hampshire some wonderful schools have been developed within the past decade; whereas in other places some public sector schools were little better than Portakabins, because educationalists failed to perceive a need for extra accommodation in time for it to be well-designed.

When we consider the British Library procurement – I will say nothing about the architecture – the fact is that it took more time to create than anyone could have imagined; its cost over-runs were manifold; it was the only major cultural building erected during Mrs Thatcher's years as Prime Minister and yet she never mentioned it in a speech. I think there is a lesson in that! We did not acquit ourselves well in the procurement of that public facility.

I think it is reasonable to conclude that traditional forms of procurement were disappointing, programmes were late and costs were often high. We had all sorts of attempts to remedy that. Design-and-build was the one which most local authorities and public procurers turned to, particularly during the 1970s. You can spot the results. Like the schools, the town halls procured by design-and-build offer little more in architectural terms than the Portakabins which they replaced.

The reasons for the Private Finance Initiative, therefore, are threefold. The poor record of procurement and control of costs; some poor buildings derived from that procurement method which have been costly in use (public sector housing is an example of that, erected mostly in the 1960s and now already at the end of its useful life); and finally, the very important need for Government to take this area of public expenditure, once paid for by taxes, outside the public expenditure budget.

At this point I want to make clear that the Private Finance Initiative is not an initiative to provide buildings. This, in part, is the nub of the problem of good design quality. It is in fact an initiative for the provision of services. The Private Finance Panel document on this gives some useful examples.

A school gymnasium is not an output and is not a PFI, but regular access
to sports facilities is a PFI. In a hospital a heating plant is not a PFI, but a
reliable supply of heat may be. An aircraft simulator is not a PFI but a
training service is. A computer system is not but an information service is.
An office building itself is not a PFI but the provision of service and office
accommodation is a PFI.

These are very important distinctions. They make one realise that what we
are concentrating on is what happens as a result of PFI. In the case of
schools, for instance, the provision of services to support the education of
children will inevitably require a building which we regard as a school;
but it is the provision of services for which we are bidding and competing.

At its very foundation, then, the Private Finance Initiative is not an
initiative to build fine buildings of architectural merit, or roads and
bridges which will rank amongst the world's best. But it could produce
those results if we are prepared to use it well and broaden its scope. I
firmly believe that that is the responsibility of PFI clients and bidders.

Who are the bidders for these PFI projects? In most of the property PFIs,
they are a strange assortment. I will tell you how the bidding consortium is
brought together. An advertisement appears in the *Official Journal of the
European Commission (OJEC)*. It is quite tricky to know what these
advertisements mean but, once you have the hang of it, you realise that
there is a building involved. That usually gets contractors, banks,
architects, specialist healthcare operators if that is appropriate, and facility
management providers, on the telephone ringing up to say 'Are you going
for this? Have you formed a consortium?'. There is a period of a week
when one dances and flirts with everybody else, until a consortium comes
together. That is what is happening now, unless one has been tracking the
likelihood of a Private Finance Initiative coming forward and knows about
it before the *OJEC* advertisement appears.

It is, you may agree, a very strange way of putting a team together. That is
the first point I want to make. Many of these bidding lists are led by
contractors. Their motive is to obtain a building contract. Others are
headed by bankers, who want to syndicate the loan, which is quite easy to
do because it is generated by an AAA covenant, being with Government;
they can sell that income stream and trade paper.

Then there are the facility management providers. The last thing that they
really want to do is to put equity into the project. They want a contract for
the supply of facility management services. Then you may have the
specialist healthcare operators; the IT and equipment technology suppliers,
who have resource and muscle; and the developers, such as Stuart Lipton,
Godfrey Bradman, British Land, Hammerson's and myself.

Who is providing the impetus? Who is the client who wants to own this
contract? I suggest that this question points to one of our problems. At the

moment we have opportunities being addressed to a market which is ill-formed to take advantage of those opportunities. The private sector must decide what constitutes a company which makes an ideal bidder for a PFI contract. That process has some important implications for the way in which architects are appointed to a project.

The Private Finance Panel's documentation gives some good advice about the appointment of advisers. It says first of all you have to have a financial adviser. They take the lead and are first on the scene, and they are very good at working out finances – and one pays the fees that go with it. Then you need a legal adviser, because there will be a huge amount of documentation, perhaps amounting to several cubic feet – which makes one realise why these two particular groups should be so very interested in PFI. Then you get the technical advisers. They include, in this order: surveyors, engineers, architects, contractors, project managers, actuaries, specialists. The architects are a very small item in that bunch, not like the legal advisers, and not like the accountants and financial advisers.

Equally it says that, in the appointment of advisers, PFI experience is very important. I can think of quite a lot of architects who would make excellent advisers to a PFI but who only just about understand what 'PFI' stands for. What they are able to do is give advice, in the way John Allan so eloquently described, about the rôle that an architect can play in the early arrangements for a PFI project.

It does not always happen that PFI projects are attended by an architect of John Allan's stature. Quite often, such architectural advice as is given is from someone in the background. I believe it should be far more up front. I believe too that, instead of just seeking the provision of the services, it would be as well to recognise the possibility that we may gain buildings of real merit – and make that a very important part of the collective procurement process.

When my Group entered the process of competing for the Customs & Excise office estate in Southampton it was described as a pathfinder project, along with the Treasury project in Whitehall and the Longbenton proposal for the DSS. We appointed a local architect, Trowbridge Steele. Unusually, Customs & Excise did not have a site, but rather a number of buildings. The competitive process was to find a means of providing suitable accommodation for them in a single building, which did not need to be on any of the existing sites. There was no real guidance as to what should happen to the residual estate. That is interesting because it is owned publicly and there should be some public responsibility for what happens to redundant public buildings. These particular buildings are not very good, and my idea is to knock them down. It might not be everybody's chosen route! If I had taken just a financial view, it might have been marginally better to have proposed to keep at least one of them.

In proposing a solution to the Customs & Excise question, we investigated no fewer than eight possible sites and prepared design and planning briefs for all eight of them, before coming to a shortlist of three, one of which was already Government property and involved an alteration of the existing building and its extension. We put in two sites in the final stages.

Our plans at the preferred bidder stage are at a scale of only 1:200. They are certainly not plans that I would expect could be properly judged if brought here to the Royal Fine Art Commission. They are sufficient for the advisers on the team to prepare proper costings and to advise Customs & Excise, within an area of risk, how much they will have to pay over 15 years. In this instance it is a project which we will own. At the end of that period, if Customs & Excise leave, we shall have an empty building – which is entirely our risk. In this respect it differs from many other PFI projects. Throughout the whole initiative Customs & Excise did have an architect – Pringle Brandon. I think they were represented at one meeting. I never had a conversation with them and I do not believe my architect did either. Architecture was not greatly in evidence as one of the concerns in winning the bid. That is unfortunate, but notwithstanding that, I hope we shall produce a good building.

At Norwich a PFI has just been announced for a police headquarters of some 15,000 square metres, twinning another one in Abingdon for Thames Valley police. It is a significant public building. Again, the trail starts with an OJEC advertisement. This is the wording:

> 'To assess and evaluate the requirements to provide a site and/or premises, finance, design, build and, at completion, provide certain non-operational support services and continuing facilities management for a county police headquarters services. The facilities to be modern, fully fitted-out offices, with ancillary accommodation totalling approximately 15,000 square metres.
>
> 'The awarding authority is looking for an innovative approach. It is open to the possibility of shared use...' etc.

Nowhere does it address the quality of architecture issue, until the next page when it says:

> 'What is required is a proactive approach which will enable it to achieve a higher-quality enhanced response.'

I think that might be about architecture. But nowhere, for a public building, does it say upfront 'We want a building that is important in the context of Norwich, which takes its proper place as a public building in an important city'. It does get a little better when you reach the pilot questionnaire, where the bidder is asked:

> 'In designing a modern police station, what would you expect to be the key design criteria to ensure that the building continues to meet user need during the lifetime of the contract?'

This question invites me to say that I think it should be an important public building, but I am not told that that is what should happen.

As I go through the documentation, I find other qualifications which give us some clue to the fact that we must satisfy the requirements of the brief:

> '...making the building capable in all respects of providing a working environment which is pleasant, environmentally friendly, energy-conserving and maintenance-minimised.'

These are all very factual needs and requirements that one is asked to fulfil. Nowhere are we asked to produce architecture of high merit, suitable for a public building.

Another advertisement which has just been published is for the Army technical support building in Andover. This one says:

> 'The requirement: provision of office accommodation for 760 staff of the Army Technical Support Agency together with catering for all staff and messing accommodation for military staff...car parking...' etc.

That could so easily have read: 'Provision of office accommodation built to a high standard of design for 760 staff' – if that were important, and I think it is important. What a small thing it would be to put upfront the question of architectural merit. It might make a difference to some in appointing the technical team.

There are other big bids which merit our concern, because they are important buildings. Chelsea Barracks is the subject of a PFI bid; so is the MOD headquarters in Whitehall, a Grade I building. This latter project includes finding new uses for three other premises, two of which will certainly fall to the bidder: Northumberland House, the Metropole and St Christopher's House in Southwark, which may lead to the redevelopment of an area on the South Bank. There are at least four projects in that PFI to which high architectural standards must be applied.

At Colchester, the Ministry of Defence are seeking bidders for the redevelopment of 316 hectares of land. It is considered by the Army that it houses the worst accommodation in the British Army. They have appointed an architect – Broadway Malyan – and I am delighted by that, because the MOD is at least taking the view that an architect should be present in the proceedings.

The National Physical Laboratory covers 80 acres on the edge of Bushy Park in West London and includes a number of listed buildings. This is a site which the DTI regard as one of the three most important world centres for scientific research. When I looked at the 1.6 cubic metres of brief which came, I went through it again and again. I saw reference to a master plan, to which we are not committed; in fact the whole brief suggests that we should look very hard at the existing masterplan and start again.

In discussing the issue with the bid promoters, I asked what they wanted and they replied that they really wanted premises which would reflect state of the art research laboratories. I said 'Couldn't you be more specific?'. They said 'We are allowing you 90 hours of interview time with the scientists, in order that you can derive the brief'.

I do not greatly object to that, but here is a very important Private Finance Initiative for public buildings of considerable importance to the nation, and the promoting body does not have a clear idea of what it wants. I will try to give them a good idea, but it is not easy to introduce into this the importance of the architectural status of the buildings.

It is not all that bad, and I want to mention three other examples which are relevant. In Pimlico there is what I believe to be a disastrous architectural experiment perpetrated by the London County Council in the 1960s. It has outlived its natural life and should be pulled down. Westminster Council agree and so do the Government bodies. Others may feel that Pimlico School is a building of architectural interest. I do not. I spent two hours there and came away more depressed than I had been in a long time. If this particular Private Finance Initiative is to work it is important to build a decent school in which young people will not just be conditioned to an environment as though they were not involved; in which teachers do not become visually immune to the ugliness of their surroundings.

The brief for the Pimlico School project says:

> 'It is expected that the architectural design and interior environment of the new school facilities will be of the highest standards and quality. The new school facilities will be expected to redress the deficiencies of the existing facilities, the overheating in the summer...' etc.

> 'The design should be responsive to its context and reflect the high quality of the school and its important place in the community.'

Very refreshing words.

> 'It is expected that the quality of materials and workmanship and the performance specifications will, like the architectural design, reflect the highest quality and seek to improve upon conventional standards.'

I think that is absolutely excellent. Attached to that was planning guidance from Westminster City Council, which had been the subject of public consultation and took effort and care in advising us of the importance to be attached to the design.

The other school is a new Jewish day school in north London. For those who think that PFI has to be complicated, let me tell you how this has gone

The former Oxford Jail, a Grade I listed building awaiting a privately-financed conversion. Trevor Osborne Property Group

forward. The familiar advertisement appeared. We put in a pre-qualification document, to receive a telephone call the next day inviting us to a meeting four days later. We attended the meeting. We were given 15 minutes to elaborate on the document. They asked us questions for 15 minutes, whereupon the chairman asked his colleagues if they were happy. They said they were. He then told us that we had pre-qualified, passed the next document to us and asked us to give them a view on the financial implications in two weeks' time. We did so and went to see them three days after the two weeks had expired. We were immediately told that we would progress to the next stage and that they would like to go for the final bids, including design, within about six weeks. I think this particular project is going to work. My professional team, including the architect, felt immediately that they had direct dialogue with the chairman of the governors and that those promoting the school were committed to the project. It is very refreshing.

Turning to historic buildings, I think it will work there too. I am bidding for the re-use of the old Oxford Prison – a largely 19th century Grade I structure which I want to turn into an hotel. It is in effect now owned by Oxfordshire County Council and should have a use which perpetuates public access. Its conversion should be well designed. This is the sort of project which can properly be run as a Private Finance Initiative, if it is directed in a way which will achieve objectives beyond the provision of service facilities.

The problem boils down to a number of factors:

- First of all there is an emphasis on value for money. How that is measured is so important. It is the intangible financial benefit of a good building, which is hard to quantify. There are some things you cannot easily put a price on, but which you know improve the quality of life.

- Second, there is what is called the public sector comparator. That is a calculation based on past experience of other procurement methods, although I would put it to you that those methods have not always produced the best buildings at the expected cost.

- The third problem is that which I mentioned earlier – the initiative is about the provision of services and not the provision of the highest quality architecture. The promoter needs to insist that buildings of high architectural merit, reflecting their public purpose and status, are provided as part of the initiative.

- Fourth, bidders for property PFIs are largely contractors and bankers, with only a transitory interest in what is achieved. I think there is a real need to see patrons of the art of architecture in the PFI system, joining with the promoters as the more distant patrons in achieving buildings which are worthwhile.

- Fifth, the link from the architect to the end user who is not his client is a distinct problem. When we are bidding for the provision of a facility, we are constrained during the important first period of design from sharing too much information with the promoter. With Pimlico School, for example, we discovered before the final bid was submitted that the proposed school was too large to meet Department of Education standards. We were foolish enough to tell the promoters that, which meant that they told all of our competitors. Such a shame! It is this sort of experience which makes one very reticent about sharing all one is learning through the design and bid process with the promoter. This sometimes makes the architect remote from the end user of the service, and interposes a client who is not the user – merely a group, a body, an organisation, which will obtain a financial return. I do not have a solution to this difficulty, but it certainly needs very careful study, so that a close bond is formed between bidder, promoter and architect in a way which will work for the architect.

The rôle of the advisers in the preparation of the bid and the pre-qualification for short-listing is very important. An architect must be appointed for any bid, in my view. It is important that the architect is given the sort of instructions that John Allan was given, and that there is a clear understanding that he may not be appointed as the architect for the building which will emerge. I believe, however, that the architect should be present and should guide those assessing the brief so that they know whether the bids they are receiving include a high standard of design. Any bidder failing to meet that required standard of design should not be considered, no matter how attractive the financial bid might be.

Finally, the cost of providing high quality submissions. It is no good the public sector believing that the private sector will shoulder the costs of bidding indefinitely and that architects will join in that gamble. It is unfair to architects and is not a good way to procure a high quality of architectural design. It is something which deserves the close attention of the Royal Fine Art Commission and indeed the Department of National Heritage, because I think it is at the root of the problem. It is quite clear that at the first stage of design architects are selling time as well as experience and talent, and they do need to be remunerated. After all, it is hard work winning these bids, although the real work is afterwards.

I think that PFI will continue, even with a change in administration. The public expenditure question means that we must make PFI work, because there is no other way to procure public facilities and buildings on the scale which is now needed. I read in the *Financial Times* that visitors are coming from China, India, South Africa, Spain, Thailand and Israel to ask how we do it. They are asking the wrong people so far. But, with a bit of effort, and if we get it right ourselves, we might be able to produce a great export opportunity for British design and know-how. It would not be the first time that we have exported those things.

PFI offers an opportunity to achieve more than merely fitness for purpose. We cannot sacrifice our future heritage to value for money, badly judged. We have to take into account other considerations so that this generation contributes its share of high quality public buildings.

Discussion

Hal Moggridge (RFAC, Past President Landscape Institute): I notice that we have reached this stage of the day without mentioning the site, except when Trevor Osborne mentioned Bushy Park. I find this very worrying, because sites are very much part of what has to be procured. Successful site development needs consideration in the context of a long timescale: 25 years, for instance, is barely enough to establish trees of significant size. This is a subject which needs some very serious analysis in considering procurement by PFI.

Robin Nicholson (Edward Cullinan Architects): The new Norwich hospital offers an example of what the Private Finance Initiative could do in the centre of our cities if it is badly managed. It is in city centres that a number of existing health buildings are being sold off, for whatever purpose, but this new hospital is being built on a greenfield site. It is entirely understandable why this is happening, because development costs are lower; but the long-term cost to the community might well be greater. It also contradicts the Secretary of State's planning guidance on the development of out-of-town projects. We have to be careful that PFI does not end up subverting the best bits of the planning process.

John Allan: I would ask Trevor Osborne whether, as a result of the 'shotgun marriages' that follow *OJEC* advertisements, he has been able to establish any groupings that have begun to gel as teams; is he pre-prepared to respond to these advertisements, as is beginning to happen in our case?

Bryan Jefferson: Partnering in PFI?

Mr Allan: Exactly.

Mr Osborne: The first thing I have learned in this context is that it is probably not the ideal solution to have a contractor as a partner. We have recently put forward one response which lists four contractors, each of whom is told that they will be in competition with the others for the three contracts on offer. We told them that they had a reasonable chance of getting a contract, but that there would be a competitive process. All found that acceptable. That approach removes the necessity of incorporating a contractor in the process. I think the contractor should have the status of a sub-contractor, competitively placed.

It is very important that the industry does begin to develop corporate organisations. It should be possible to keep a large part of our Pimlico School team together for future bids in the education sector.

Francis Golding (Secretary, RFAC): One of the things that has come out of the discussion today is that there is a tremendous need for expertise at the level of the client, the procurer. That was one very strong message I had from John Allan's presentation; the procurers had thought very deeply about what they wanted and how they might get it. Out of Trevor Osborne's presentation one could begin to see the need for potential bidders to group themselves and develop an expertise in putting bids together, but the need for the client to be trained now seems to me to be just as great.

Until today, many people had been thinking in terms of guidance as meaning bits of paper telling people how to conduct a PFI project. One of the needs which has manifested itself very strongly in my mind today is to find some better way of building the expertise both of potential procurers and of potential bidders.

The Royal Armouries Museum, Leeds

Guy Wilson, Master of the Armouries
Professor Derek Walker, Derek Walker Associates

Guy Wilson

I start by echoing what Trevor Osborne said about the fact that there are good and bad things about both public and private procurement.

The Royal Armouries project is an oddball, because we are what I would call a pre-PFI initiative. By the time the Treasury issued its first public document on PFI we were already out to investors with a placing document. We started off in the public sector and ended up in the private sector. It was the decision of the Royal Armouries to seek a way of increasing the size of the museum by moving a substantial part of it out of London. All we knew at the outset was that the Government was not going to find all the money. So we began as a public sector project, knowing that some private money would be required.

We gradually moved the process forward stage by stage, showing that it could work. All the time we were looking to draw investment, both from the public sector and the private sector. We started off by saying 'We have a timescale in which to do this, because we are running out of money and the Government is running out of money. It will not get any better. If we want to do this project, we have to do it quickly'.

We were also aware that in the late 20th century an arms and armoury museum, even if it is the oldest museum in the country, will not get large amounts of privately donated money, and it was highly unlikely that any public appeal would do anything other than drag out the process.

We started to look at what our assets were and what we could offer people. The asset we had in the Tower of London was that we were in valuable real estate, out of some of which we were prepared to move – and there had to be value in that. That is what we had to persuade central Government of in order to get the £20m eventually offered to us by David Mellor. We then had to go to the cities – the places we were looking to put our museum – and see how much money they would give to build up a

The Royal Armouries Museum in Leeds, showing the octagonal Hall of Steel overlooking the River Aire. A spiral staircase allows visitors to circulate between floors and view exhibits within the tower's concrete core. Derek Walker Associates

package, so that we could eventually go to the private sector and say 'We are putting up "x", in return for you raising "y". These are the benefits you will get as the operator, working with us for the next "z" number of years'.

The site was vitally important. That again was our decision in the public sector. We eventually sought bids from a number of cities in the north. Deciding to go north was not entirely altruistic, although there was an

element of altruism involved – thinking that we should be doing
something to help regenerate an inner city. We were also aware of things
like development corporations. We took bids from six cities and put
together a range of criteria for deciding which city and, within the city,
which site, should be chosen. In the event, every member of the board
chose Leeds. It was the one city which met all the criteria we had laid
down and which had the best possible site for our purposes.

We put together the consultant team and appointed the architect. We
decided that we would not go down the architectural competition route,
but would go to a limited number of architects and seek someone with
whom we could work. We did that for a number of reasons. We knew that
we were not particularly skilled in developing new museums. None of us
had ever done it before. How therefore could we properly judge the results
of an architectural competition, as far as the work of the museum was
concerned? We were also aware that though the architectural profession
had far more knowledge of museum design than we had, none of them
had experience of designing an arms and armour museum in the late 20th
century to our specifications.

We wanted to develop the museum from the inside out, starting with our
collections. After we selected Derek Walker Associates as architects, we
had a long gestation period during which a very small brief developed into
the final, very large brief which dictated the design of the building. Derek
and the other consultants were all appointed by us, knowing that they
might well be novated across to the private sector in due course.

We had to put a completely costed scheme together before we could sell
the whole idea to the private sector. The private sector had never before
been asked to look at a private sector museum or heritage attraction. We
had literally to spoon-feed them.

We had to decide what sort of building contract to use. One of the first
people we employed was a financial consultant to work out how we could
put together our business plan and go out to the private sector to attract
investment. We were told at an early stage that the private sector would
accept only one risk on this project – the visitor numbers when they were
operating it. They would not accept anything else, so we would have to
guarantee them the capital cost of the project. Anybody involved in these
things will know that you cannot guarantee the display cost, because you
work it out as you go along. All you can do is put a cap on that and say
that once you have spent a certain amount of money you will stop.
Therefore we had to guarantee the cost of the building. In the end we
opted for a design-and-build approach with a greater than usual element
of design, which we called develop-and-construct.

Our financial consultants discussed with the Treasury the most
appropriate form of partnership with the private sector. Seven or eight
different ideas of cross-partnership were put forward, and we came up

with one where basically we hold the land and give a company money to
build the building we have designed, on our land, and operate it for a
certain amount of time; at the end of that time the building reverts to us.
The private sector operator takes a profit for 'x' number of years – it turned
out to be 60 years – and the project agreement then said what, within that,
the Royal Armouries as a museum must do to preserve its statutory
responsibilities for displaying the collection. We phrased it as the duties
and responsibilities we would provide to the private sector on an exclusive
basis, for which they had to pay.

The Royal Fine Art Commission asked at the time whether we could
guarantee the architectural quality, both before and after the building
had been handed over. That raises two major issues of quality when
involving the private sector in this way. Can they deliver the capital side
of the programme – in other words, build the building – to an acceptable
quality? Can they then operate what they have built to an acceptable
quality? PFI and even our pre-PFI is not just about building but about
operating as well.

At this seminar we are dealing with the first of those two questions, but it
is important to understand that there may be different skills required to
deliver the first and the second requirements. Perhaps we sometimes need
to consider different people in different organisations to do these two
things.

Looking at the first question, the quality in the building, there were two
linked issues as far as we were concerned: cost and organisation. Cost was
a critical factor, partly because we entirely misunderstood the cost of all
the necessary legal activity and the cost of the cross-warranties demanded
by the private sector – which took a considerable amount of capital out of
the project. The private sector had to have the organisational ability to cope
with those problems while delivering the project on a very fast track.

The building contract was signed on 14 December 1993; we started on the
site in January 1994; the footings were laid in March 1994; the Queen
opened the building in the middle of March 1996. It was a very fast
process, therefore, and a phenomenal achievement by everyone involved.

Professor Walker

In following Guy Wilson, I would like to dwell on the reservations and
ambivalence I feel towards this particular PFI scheme.

As Guy said, it was a a PFI evolved rather laboriously before PFI had been
formally adopted as a serious policy by the Government for the
procurement of public buildings – and in response to Trevor Osborne's
comment that many participants did not know what PFI meant, I can say

The Hall of Steel provides a dramatic exhibition space. Derek Walker Associates

that after this particular experience there were moments when I thought it should have stood for Potential Failure Incorporated. However, it was the vehicle which allowed us to build the Royal Armouries and for that both Guy Wilson and I are particularly grateful.

I was also reassured to hear Trevor Osborne's plea against the use of design and build with PFI initiatives. I am afraid that many investors, entrepreneurs, banks and indeed many consortia selected for PFI initiatives seek to use this route. Time and money are critical to the success or failure of any investor's initiative. What PFI needs in order to achieve an elegant solution is a perfect business plan, perfect personnel (client and user, consultants and eventually contractor), perfect documentation and a perfect contract.

We must also realise that a chief executive who can oversee a building process is often not the kind of animal to manage a finished facility. There is often the problem of profound cultural differences between user and provider which can lead to immense difficulties in operation. It is not just two separate cultures, it is often separate agendas fuelling the flames of discord.

At the Royal Armouries, we still had problems despite taking documentation to RIBA stage E and sometimes stage F, producing comprehensive room data sheets and running a conventional contract for the museum displays. We were asked to stay on as client advisers, rather than being novated to the chosen contractor. We therefore had no legal position within the contract to ensure that high standards would be upheld and that the contractor would comply with the drawn aspirations. The only course left to us was to try to sustain quality through forcefulness, bloody-mindedness and personality rather than through legal means – by chivvying, checking and monitoring all drawings and by being thoroughly unpleasant. This ad hoc method allowed us to uphold about 85% of the building's integrity and intention.

In PFI, therefore, the refinement of the documentation and the partnership of the people carrying it out is the most important single factor. I do not think PFI will ever work if it goes on different routes and tries to involve other kinds of procurement. It has to be, in the purest sense, the perfect team to produce perfect results.

Perhaps I could now consider the building itself. Without doubt we have in the Royal Armouries one of the greatest collections in the world – a creation of extreme beauty which required a setting of the same quality. In the event we had a site which was far from perfect – an inner city site, an area of industrial blight. We located the site for the main building on a promontory, with the building looking across the river and also towards the dock. A master plan was required before we could embark on detailed designs.

The building had to be an icon for the city and a dramatic beacon for the visitor. It also had to be animated, because we knew that in the first five or six years there would be very little else on the site. An internal street would attract the public and offer restaurants, shopping and also the facilities of the temporary gallery and education department.

Tremendous care was taken in the selection of materials for the museum. We chose metallic brick, used primarily because it goes silver and then very dark grey in the rain, so hinting at the contents inside and the quality of the plate armour. The granite on the base of the building led to a very

The central atrium street leads visitors through the museum. Derek Walker Associates

simple panel of colours overlooking the waterways. Within the building we tried to use materials which were simple but sturdy and hard-wearing.

As I have said, the display contract operated as a conventional contract, with several people tendering for individual elements. The contract value was £7m, which included over £1m for making 42 films. We had tremendous value for money. This contrasts with the quality of workmanship on certain elements of the design and build shell, although when you consider that the contract for 270,000 square feet was only £23m, you can sympathise with the main contractor to a degree. That figure was really too low for a national museum of high quality. Fortunately the nuances of disappointing construction are not evident to the visiting public.

In summary, is a national museum a suitable candidate for PFI and all the baggage that goes with it? The answer in our case must be no. It is particularly rewarding that we brought in the project on time and on cost, but what was the penalty? A profoundly flawed business plan – a legal minefield with cost premiums to match – meant that the building budget was reduced by nearly £3m at inception. A similar reduction on the display budget meant that Guy Wilson and I had to lower our sights,

particularly in the quality of the hardware and some of the interactive
displays. In essence, we had to run to stay still, plug the gaps and hope
that certain disappointing omissions did not affect the visitor.
It was a wonderful adventure and though I prefer a conventional contract,
in the end I know PFI is here to stay. It must, however, address only
suitable building types, be scrupulously manned and bring all parties on
board at the outset. It is like every other initiative – only the quality of the
performers makes the performance memorable.

Guy Wilson

Our experience is that PFI can deliver quality. A partnership between the
public and private sector can bring together different areas of expertise. It
is possible in the public sector to deliver on time and in budget, but it does
not happen very often. PFI and the involvement of the private sector does
help that; though there were certainly times during the Royal Armouries
project when all of us involved wished we could slow it down, stop, wait
for six months and then carry on. We could not, because the private sector
had to have its income rolling in.

As Derek has said, success with PFI depends on how you do it and who
you do it with. It is certainly not a panacea. It can cost more money in
certain circumstances. For instance, the public sector does not have all the
insurance needs of the private sector. PFI can also cause problems for
management and cost control. As Derek says, there are elements of the
Royal Armouries Museum which are substandard and which will cause
problems in the future. That will perhaps impose a greater maintenance
burden upon our private sector partners than they imagine. Their
responsibility now is to manage and operate the museum commercially,
working with us. But they are the ones with the money. We can put in
temporary exhibitions and help around the edges, but the building is now
their responsibility for the next 60 years.

The jury is still out on whether a commercial company can run a national
museum properly over that timescale. So successful is the corporate
hospitality in the museum that our private sector partners already want to
make major changes to the Royal Armouries Hall, which is also the main
temporary exhibition hall in the museum. Here we have an operating issue
which could affect the quality of the building. If we have got the other part
of the PFI arrangements right – the project agreement – we, as landlord,
should be able to say yes or no to certain things, thus retaining a
reasonable control over the quality of the environment whilst allowing our
private partners freedom to operate commercially. We shall see.

No one said it was going to be easy, and it is not easy. Someone said to me
that 'hell was getting all you want'. That is quite right, but I prefer to be in
this hell than in the hell we used to be in. As I said at the beginning, I will
be eternally grateful to PFI for the opportunities it has given us.

Overview

Alastair Ross Goobey
Chairman, Private Finance Panel

My colleagues tell me that the day has been full of insights and animated discussion. I can neither promise succinctly to summarise what has been said, since I was not here, nor to answer all the points which have been raised; but what I want to do is to show you where we are in the PFI in relation to the topics today – how design quality can and should be sought and achieved within the PFI process.

The point I would make above all is that without PFI the prospect for getting new, large buildings in any part of the public sector is much reduced. We have to accept – and I think all the political parties have accepted – that PFI is the obvious way in which most of these additional buildings, and even some replacement buildings, will be procured.

PFI represents a remarkable change, the biggest change in public sector procurement in a lifetime. It has concentrated people's minds on the various elements of what we could call the Design, Finance, Construct and Operate (DFCO) process. The questions of design, fitness for use and the costs of operation in the long term surely should have been considered as part of the traditional public sector procurement process; but in many cases they were not. Nothing was gold-plated in the traditional process and it is no good expecting suddenly to have everything gold-plated under PFI.

One of the problems with old-style procurement is that the contractor is not very concerned about what the building will look like in 20 years' time. What Guy Walker said about the Royal Armouries Museum is absolutely true: if there is a problem in maintenance it is a problem for the private sector, which is probably where the risk ought to be. This is the essence of PFI: it has started to allocate risks where they are most suitably left. It is accepted, of course, that the public sector has a responsibility to leave the publicly built environment in a reasonable state. It cannot transfer all that risk to the private sector. But there can certainly be no defensible reason for the built assets enshrined within a PFI contract to be of poor or inappropriate design quality. It simply does not make sense for design not to take account of the needs of the end user.

It is clearly important that the designer and the operator come together at an early stage in a PFI project, and that the designer has continuous input throughout the procurement process, because it is the operator who will bear the costs of the design for up to 60 years. In many cases the asset will stay in the hands of the private sector beyond the life of the PFI project, so the private sector has a great deal of interest in the quality and longevity of the buildings.

Despite that, we hear frequent criticism that PFI is just design-and-build by another name, that it aims for the cheapest buildings, that it further distances the architect from his client and so on. I know that there has been an enormous amount of frustration from the private and public sectors about PFI over the last four years, but I do not think any of those criticisms are valid. In particular, I do not accept that PFI further distances the architect from his client. The designer's real client under PFI is the operator/end user, because under PFI the client for the likely life of the building will be the operator. The logic of PFI is that it should bring the designer and the operator closer together. The procurer should be able to conduct or be party to the all-important discussions between the two.

The Private Finance Panel has no wish to pour its energies into promoting an initiative which makes a negative contribution to the overall quality of the public realm and the public estate. I have to point out here that the Private Finance Panel is one of these wonderful British inventions which has no power whatsoever. All we can do is try to chivvy and persuade. The only thing we can really get ourselves involved in is to try to clarify the PFI process and to act as intermediaries between those who are trying to procure things in the public sector and those who are trying to build or design them.

Our only remit is to make PFI succeed. Defining success is more difficult, but we would certainly include demolition of Marsham Street as a good example. In the interests of success, we do listen to what we are told. The views expressed, the issues raised and the experience shared here today are all of great value to us. We will try to learn from them, and also from the exercise we are engaged in with the Department of National Heritage.

What has come over loud and clear, not only today but on other occasions, is that there is no set recipe for PFI schemes. There is no sequence of procedures or contract clauses that can be followed in order to achieve a satisfactory result. That is one of the great frustrations for everyone involved. The element of how much risk is transferred, where the risk is transferred and the length of the contract tend to be different from project to project. And of course we are not going to build another Royal Armouries Museum, so the experience of that scheme is not directly transferable. The process is transferable, but the actual project cannot be replicated.

What also came over loud and clear is that good design quality and PFI are not mutually exclusive. It is possible for the public sector procurer, at the very start of the procurement process, to set an initial design framework.

That may, of course, carry dangers, in that if the public sector procurers keep to their traditional framework, then the innovative potential of PFI may be lost. When prisons, for example, were put out to PFI tender, there were some very innovative ideas put forward, including prisons without walls and prisons with moats. The design of prisons has not changed much since Jeremy Bentham designed them in the 18th century, so some innovation in design is not to be sniffed at. We hope PFI gives the architectural and design professions some incentive to come up with some new ideas about how to address the problem of delivering services. The procurer should try to make design considerations one of the criteria for evaluating a project.

The experiences which have been shared here today offer certain pointers to prospective PFI procurers about which routes are most likely to promote good design. One moral is that the initial formative relationship between the designer and operator is very important. Finance and construction come along later – though not much later, I hope. One of the problems we have found with PFI is that the finance has come in too late and the whole project has had to be rejigged because the financiers would not finance it in the way which had been assumed.

One of the Panel's duties is the instigation and dissemination of good and best practice in PFI procurement. Some of you may be familiar with the many guidance notes the Panel and the Treasury have written and published. These are aimed primarily, but not exclusively, at those in the public sector trying to get PFI procurement into practice. We want to be sure that they will be able to get the best value for the public purse and the public sector user.

We have now joined forces with the Department of National Heritage to draw up and publish guidance which will help procurers define their projects and proceed through every step of the PFI procurement process, in those ways which have the greatest likelihood of securing the best quality of design. I will not trail the content of that guidance note, because research, canvassing of views, winnowing and refining are still in their relatively early stages. But it does show you that the Panel is committed to finding ways of encouraging the proper and timely incorporation of design advice and design evaluation into the PFI process, in order to achieve well-designed facilities. In the context of the forthcoming General Election, I would add that we think our guidance will be of assistance irrespective of the nature of the new administration.

Please do think of the Private Finance Panel as a potential ally in the fight to get good design in the PFI. Any good ideas are always gratefully received by the Panel. We are trying to make sure that people think of design at an early stage and that it will not be ignored in the procurement of PFI projects. Not everybody will be happy with all the results but at least you will feel that design considerations have been given their proper place.

Discussion

John Lynch (Nottingham City Council): Looking ahead 20 years, when the PFI projects now being generated have run their course, presumably the relationship between the public service provider and the private sector partners (who by then possess the asset) will need to be renegotiated. The revenue commitments of the public sector could become quite damaging.

Alastair Ross Goobey: One of the problems of traditional public sector procurement is that we have often ended up with assets which we do not want in 20 or 30 years' time. The life expectancy of many assets, in terms of their usefulness, has declined dramatically over the last 30 years. The public sector has been notorious for taking out over-long leases, or for buying buildings which within 20 years have become entirely inappropriate. That is one of the problems PFI is trying to avoid.

If the public sector still wishes to deliver the service from that particular asset after the contract period, then it will have to negotiate rents with the owner of the asset; but (perhaps because of technological change) it might have decided by that stage that the service it wants to provide is no longer best provided in that asset.

It is certainly true that Government departments are very anxious about over-committing their budgets. That is a problem for the politicians because ultimately it would have to be faced anyway under conventional procurement – except that they would then have a real cost going out, which they probably have not recognised up till now. It is up to the politicians to decide what level of public expenditure is bearable.

Trevor Osborne: Some of the Private Finance Initiatives do pass the ownership of the asset to the private sector. Then there is a contract for the use of that asset for a period of years, at the end of which the bidder, the provider of the service, owns the building. It is then up to the public body to decide whether to renew. There will certainly be difficulties relating to the employment of people providing the service at that stage, but those will take the course of most legal contracts.

As Alastair Ross Goobey said, we are seeing the public sector wanting to give up many of the buildings they have, because they are worn out or they have no further use for them. They are quite willing to pass over the freeholds and the leases and let the private sector deal with them financially as best they can.

In some initiatives, for instance schools, it looks to me as if 25 years is the sort of period that the private sector will sign up for. At the end of that period the school belongs to the promoting authority – the local education authority if it is a school.

One of the comments David Steeds made this morning was about designing buildings for a short life, because technology – certainly in healthcare – might mean that the building is redundant after 30 years. The way PFI seems to be working is that if you have a 30-year contract you amortise the whole cost over the 30 years, so that the public body running the facility will have completely paid for the building during that period, as well as the capital interest on it. That is another good reason why you should go for higher standards of design and make sure that the buildings are maintained, because they are free of charge after the initial contract period.

Norman Rose (Business Services Association): Trevor Osborne talked in his presentation about getting together a slightly more solid grouping which did more than one deal. One of the issues we are looking at in the service industry is having agreed groups who know each other and who therefore will come to a new bid knowing the strength of each party. This will speed up the bidding process and encourage a better product at the end.

We may have three or four different consortia groupings which we can bring together at different times for different projects, as appears best to them and to us. That may well be a way forward which helps the designer and the operator to come together quickly and to cooperate with the construction company and the financing company.

My second point is rather more general. Today we have had a very clear view that designing and operating go together. Coming from the operating side as I do, I believe there is no better body to take this forward than the Royal Fine Art Commission. My Association would certainly be very happy to cooperate in that, and I am sure the Panel and the DNH would also be interested. It is an area where the Commission can make a unique contribution to taking forward high quality in the Private Finance Initiative.

Michael Gwilliam (Civic Trust): First, could we be given the timescale for the guidance mentioned by Alastair Ross Goobey? It seems to me that there is a need for that to come out relatively quickly.

Secondly, to what extent is there a rôle envisaged for PFI in the area of housing?

Thirdly, on the issue of the life of buildings and the use of resources, I find a certain tension between saying 'let's pay for it in 30 years and then let it go' and the pleas from other quarters about conserving resources in the construction process. We need to take a careful look at the issues of whole-life cost and the use of resources.

Bryan Jefferson: The first of your three points related to the timing of the working of the panel which has now been set up jointly by the Department of National Heritage and the Private Finance Panel to give some guidance on PFI design. Our hope is that we shall have something available in the early autumn. We are charting unknown waters, so do not hold me too rigidly to that. We want everybody associated with the guidance to feel an ownership and to support it, so that it will carry considerable authority.

Your second point about housing is interesting. It is perhaps significant that housing *per se* has not featured at all today. I know that collective housing – student accommodation, for example – certainly comes within the PFI context.

Michael Gwilliam: It is housing association developments I had in mind.

John Allan: I am involved in one. It is a housing association response to a local authority initiative.

Trevor Osborne: If there were a change of Government in May I would expect initiatives to stimulate the use of PFI in the financing of housing.

Bryan Jefferson: The final point relates to what may have been a slightly misleading impression of the tenor of the day from the most recent exchanges. I hope it does not lead to a feeling in the back of people's minds that these buildings have been built over an operating contract of 25 years, only then to be thrown away without careful consideration. Indeed, the view has been expressed that DFCO is more likely to produce buildings which are cost effective in use and have an efficient design life, and it may be that interest in sustainability will produce a solution for their use well beyond the contract life.

Alastair Ross Goobey: An operator who is left with a building can presumably do what he likes with it. If it is a well-built building there will be more things he can do with it than if it is going to fall down.

Michael Gwilliam: I can see that, as long as it is not seen, as it could be seen by some people, as an 'easy in/easy out' option. The other point is that somebody is presumably paying quite a lot in that 25 years to amortise the cost.

John Wells-Thorpe: May I add a rider to Alastair Ross Goobey's comments? He referred to the fact that in his view there was not a gulf between the designer and the user. I beg to differ on that.

One of the jobs of the designer, if he is brought in early enough, is to act as devil's advocate. If someone says 'We want a new district general hospital', it is for him to say 'But are you sure? Are you convinced that you do not see the shift between the secondary and primary going in different ways?'. The designer cannot offer that quality of analysis if he comes in as part of the construction team. So I think there is a gap there. It is a gap which can be made less wide, but there is a fundamental disadvantage in the system as it is practised at the moment.

The other thing that worries me is the notion of an ever-decreasing life in the structure itself. The history of post-war prefabs, some of which are now listed, tells us that you do not get rid of buildings at some theoretical date. The problem about having a background notion in your mind that a building is only there for 25 years is that it will inevitably lower standards of specification. The whole notion of quality, which is essential to our debate today, will be under even greater threat.

The old notion of 'long life, loose fit, low energy' still prevails for many of our public buildings. One can think of endless numbers of buildings whose uses have changed over millennia. Diocletian's palace at Split, for example, has been everything from a fortification, palace, prison, hospital and tourist centre, and it goes on being used, because it has that loose-fit quality to it. I do not think we should lose that.

There are some things you can throw away after 25 years. I am perfectly happy to accept that. But the notion of a building within a loose-fit envelope, whose external appearance is sympathetic to the environment in which it sits, is an equally valid way of approaching public investment.

Michael Gwilliam: I would like to support that last point. There is a problem with the notion of a fairly tight fit – which, from all I have heard today, is where the process is tending to lead – because of the understandable need to be as economical as possible in the process. It will be need to be quite a creative process to create a loose fit within that context of resource constraints.

Alastair Ross Goobey: All I would re-emphasise is that the difference between PFI procurement and traditional public sector procurement is that in the former the risk of obsolescence in a building is transferred to the private sector. If the private sector would like to be able to use the building after the initial contract is over, it is in their interest to have a loose-fit building.

As Mr Wells-Thorpe rightly said, however, the whole process of delivering healthcare in 25 years' time may have changed dramatically. Traditionally, the Health Service ended up with a hospital which was no good to them – and we have plenty of old hospitals around as a result. My local hospital is Bart's. Everybody is pleading for Bart's to be kept, but I cannot think of a less appropriate building for a modern hospital. The most modern part is the mortuary and the second most modern part is the canteen block; but the hospital itself is hundreds of years old. However, the public sector has it. It can do what was done to Diocletian's palace – perhaps turn it into a Museum of British History.

John Allan: I want to come back to the question of risk and ask Mr Ross Goobey to say something about the Panel's attitude to at-risk work, as a factor which is now quite a serious ingredient in the disenchantment which exists in my profession and others.

Bryan Jefferson: You are not talking about the risk inherent in a PFI project but the risk associated with the pre-bid acceptance schedules?

John Allan: That is right, and specifically the relationship of risk to reward and the question of gearing, which is so different depending on which bit of the process you are participating in.

Alastair Ross Goobey: This question about the disenchantment of those of you who have had to invest a lot of time, money and resources into trying to get into PFI projects is one with which we are painfully familiar – particularly in health, because we do not yet have a health project off the ground.

In some other areas we are beginning to see a sufficient number of deals coming through to enable risk capital to be recycled. There is a lot of discussion within the public sector about the lessons we have learned from the projects which have come through. Is it right, for instance, to get to preferred bidder stage quickly and then spend years negotiating with the preferred bidder and spending a lot of their resources? Or is it better to have two competing preferred bidders, maybe even reimbursing the losing bidder for their costs on the basis that you get competition in price up to the final contract stage? These are exactly the sorts of questions the Panel is there to wrestle with.

We are only four years into it and are still learning. That really is not a long time in terms of the way we are changing procurement. Please keep shouting and moaning, by all means. We are trying to refine the process.

Bryan Jefferson: I am glad you said that, Mr Ross Goobey. I am sure many people here will draw encouragement from that.

It says on your programme 'Chairman's summing-up'. I have taken six pages of notes and, you will be glad to hear, I shall disregard them all. I have four points which have come through to me as having an underlying influence on what we have been talking about. I want to try them on you. If you do not agree, then can I invite you to write and tell us at the DNH/Private Finance Panel steering group?

First, design quality generally and a couple of underlying messages which remain valid whether it is PFI or not PFI. The first is the need for a well-considered brief and the value of the interface between design advice and the client, whoever that may be. A client cannot absolve himself of responsibility for the quality of the building that will finally emerge. Too often important strategic decisions are taken on the nod by busy people who unwittingly, by that decision, set up constraints which will impair permanently the likelihood of that building succeeding.

The second general point on procurement is that adequate time and resource, and that means money and talent, must be put into the inception stages. Attempts to cut back on time or resource will have a detrimental effect, from which it may be impossible to recover.

On PFI, three general messages:

- The first step must surely be an early examination of the suitability of a project to PFI and which, within the PFI procedures, is the best one. David Steeds said this morning that he did not expect more than about 20% of Government procurement to follow the PFI route. Which 20%, and why? Let us look long and hard at suitability.
- Second, early expert advice on design-related matters must be readily available to the procurer. That advice, having been achieved early, must be available throughout all stages of the process. You cannot tell the

procurer's adviser, once some sort of draft outline has been prepared, that he can go home. He will be necessary to vet, monitor and help to secure design standards right through the project.

- Third, preliminary design work, properly briefed and resourced, must be done before bids are accepted. A linked comment is that in maintaining the competitive process through what is sometimes a very long period, we have to recognise the risk, the duplication and the speculative nature of much design work. If high quality is to be achieved, these costs somehow must be considered. Otherwise, that early design will not be of the quality to secure a satisfactory product.

In my book, good design and PFI procedures are not mutually exclusive. PFI can work, but the importance of design quality must be built in if we are to succeed in producing a generation of public buildings of which we can all be proud.

That is the conclusion I have drawn from a very fascinating day. I want to end with three words of thanks. First, to all our speakers. They have all had a round of applause and they have fully deserved it. I would also say how much the organisers appreciate the time and trouble people have taken to come and give their views. It has been a marvellously balanced presentation of the different perspectives.

Secondly, I would thank all the delegates, who have been very attentive and forthcoming in conversation. Finally, I thank the Royal Fine Art Commission for having the perception to set up the day, and the ability to organise it as well as they have.

Seminar delegates

Chairman: Bryan Jefferson CB CBE
Stephen BARTER (Richard Ellis)
Richard BURTON (ABK Architects)
Arnold BUTLER (Property Advisers to the Civil Estate)
Tony CATCHPOLE (Capital Planning)
Richard COLEMAN (Richard Coleman Consultancy)
Ian COOPER (HM Treasury)
Edward CULLINAN (Edward Cullinan Architects)
Alasdair EVANS (Private Finance Panel)
Alison FRENCH (HM Customs & Excise)
Peter GIDMAN (Department of Transport)
Richard GIRLING (Norfolk Wordfarm)
Francis GOLDING (RFAC)
Michael GWILLIAM (Civic Trust)
Simon HIPPERSON (Tarmac)
Robert HORNER (NHS Estates)
Michael JEFFRIES (WS Atkins)
Michael KEATINGE (Department of National Heritage)
Lynette LEONG (Jones Lang Wootton)
Christopher LIDDLE (HLM Design)
Stuart LIPTON (Stanhope)
John LYNCH (Nottingham City Council)
Caroline McGHIE (Norfolk Wordfarm)
Professor Margaret MacKEITH (RFAC)
Clare MELHUISH (Building Design)
Robin NICHOLSON (Edward Cullinan Architects)
Chris O'BOYLE (Royal Armouries International)
Tim QUICK (Foster & Partners)
George RENNIE (Derby City Council)
Norman ROSE (Business Services Association)
Peter STEWART (RFAC)
Nigel THOMPSON (Ove Arup & Partners)
Kingsley UGWUANYA (National Heritage Memorial Fund)
John WELLS-THORPE (South Downs Health NHS Trust)
Michael WILFORD (Michael Wilford & Partners)
Dr Giles WORSLEY (RFAC)

Published by Thomas Telford Publishing, Thomas Telford Services Ltd, 1 Heron Quay, London E14 4JD

First published 1997

Distributors for Thomas Telford books are
USA: American Society of Civil Engineers, Publications Sales Department, 345 East 47th Street, New York, NY 10017-2398
Japan: Maruzen Co. Ltd, Book Department, 3-10 Nihonbashi 2-chome, Chuo-ku, Tokyo 103
Australia: DA Books and Journals, 648 Whitehorse Road, Mitcham 3132, Victoria

A catalogue record for this book is available from the British Library

ISBN: 0 7277 2622 6

Printed in Great Britain by Galliard (Printers) Ltd, Great Yarmouth.